T0297351

Heterogeneous Nanocomposite-Photocatalysis for Water Purification

Heterogeneous Nanocomposite-Photocatalysis for Water Purification

Rajendra C. Pawar and Caroline Sunyong Lee
Department of Materials Engineering, Hanyang University,
South Korea

AMSTERDAM • BOSTON • HEIDELBERG • LONDON
NEW YORK • OXFORD • PARIS • SAN DIEGO
SAN FRANCISCO • SINGAPORE • SYDNEY • TOKYO
ELSEVIER   William Andrew is an imprint of Elsevier

William Andrew is an imprint of Elsevier
225 Wyman Street, Waltham, MA 02451, USA
The Boulevard, Langford Lane, Kidlington, Oxford, OX5 1GB, UK

ISBN: 978-0-323-39310-2

Library of Congress Cataloging-in-Publication Data
A catalog record for this book is available from the Library of Congress

British Library Cataloguing-in-Publication Data
A catalogue record for this book is available from the British Library

For information on all William Andrew publications
visit our website at http://store.elsevier.com/

This book has been manufactured using Print On Demand technology. Each copy is produced to order and is limited to black ink. The online version of this book will show color figures where appropriate.

Working together
to grow libraries in
developing countries

www.elsevier.com • www.bookaid.org

DEDICATION

*One of authors Rajendra C. Pawar would like to
dedicate this book to beloved parents (Bhavu and Vaji)*

CONTENTS

The global growth of industrial activity in recent decades has considerably resulted in serious impacts on the environment, polluting air and water, generating large amounts of waste, and threatening environmental sustainability. In particular, deteriorations in water quality and quantity today are a major concern facing almost all countries. A number of reports have examined the development of cost-effective and viable techniques for remediation and purification of polluted water, including chemical treatments, boiling, sedimentation, and filtration. However, these techniques do not fulfill water quality standards and are not sufficient for purification because of the large quantities of polluted water. Photocatalytic degradation of water in the presence of nanocomposites based on heterogeneous semiconductor has been shown to have significant potential as a low-cost and environmentally sustainable technology for water purification. It has several advantages over conventional methods, including the ability to degrade a wide range of toxic compounds and to exploit solar energy and transfer pollutants from one medium to another; furthermore, it is applicable to gaseous and aqueous treatments, and requires a relatively short process time. Numerous reports have discussed the design and development of a variety of photocatalysts for use in air or water purification applications. Primarily, combinations of metal/metal oxide nanostructures (including Au, Ag, Pt, TiO_2, ZnO, SnO_2, WO_3, Nb_2O_5, $BiVO_4$, and Fe_2O_3) and conducting polymers (polyaniline, polypyrrole, and polythiophene) have been investigated for the degradation of organic pollutants and carcinogenic materials [6]. These photocatalysts have shown promising photocatalytic activity toward the degradation of a wide range of organic pollutants under irradiation with ultraviolet (UV) and visible light. Nevertheless, these photocatalysts are not environmentally benign, and they suffer from efficiency issues, limiting their practical use.

Therefore, in this Elsevier book, we explain about the pro and cons of nanocomposite photocatalysts. Further, we discuss about various challenges ahead to develop water purification devices for practical applications based on nanostructured materials. A number of approaches have been reported in the past to address the limitations of semiconductor photocatalysts. This book is divided into four chapters. In Chapter 1, we discuss about the principles of photocatalysis and electrochemistry underlying photocatalysis, including the concept of photoelectrochemical cells, reaction kinetics, semiconductor/semiconductor heterojunction, metal/semiconductor junctions, and reactive species generated during photodegradation. This chapter will help to understand the basic mechanism of photocatalytic process to new colleagues in this field. Chapter 2 covers the properties of nanostructured materials and their application in photocatalysis. Semiconductor nanostructures have shown outstanding properties, such as high surface area, high optical absorbance, supermagnetism, and quantum size effect, which widens the absorbance range for its use in water purification. Chapter 3 covers the chemical synthesis of different nanocomposites based on ZnO, graphene, Fe_2O_3, CNTs, and g-C_3N_4 materials, which are fabricated in our research laboratory and tested toward reduction of Cr(VI) as well as degradation of organic dyes under UV and VIS irradiations. Finally, in Chapter 4, the authors have given future directions and perspectives to design superior nanocomposite photocatalysts for commercial application of water purification using cost-effective techniques.

Rajendra C. Pawar and Caroline S. Lee

*Department of Materials Engineering,
Hanyang University, Ansan 426-791,
Gyeonggi-do, South Korea*

ACKNOWLEDGMENTS

This research was supported by Nano Material Technology Development Program through the National Research Foundation of Korea (NRF) funded by the Ministry of Science, ICT, and Future Planning (2009−0082580). We also thank to Human Resources Development program (No. 20124030200130) of the Korea Institute of Energy Technology Evaluation and Planning (KETEP) grant funded by the Korea government Ministry of Trade, Industry and Energy, NRF grant funded by the Korea government (MEST) (No. 2013R1A1A2074605) and by the Energy Efficiency & Resources Core Technology Program of the KETEP, granted financial resource from the Ministry of Trade, Industry & Energy, Republic of Korea. (No. 20142020103730).

CHAPTER 1

Basics of Photocatalysis

1.1 INTRODUCTION OF HETEROGENEOUS PHOTOCATALYSIS

Worldwide industrial revolution in the 21st century brought a wide spectrum of problems, mainly contamination of water with harmful and waste materials, leading to significant adverse effect on the environment and wildlife. Direct disposal of industrial compounds into water makes it unsuitable for drinking and for other purposes. Particularly, nonbiodegradable and undesirable chemicals have negative consequences on health of humans and aquatic life. It was found that millions of people have no or less access to pure water, and a few million people have died every year due to diseases caused by polluted water [1]. This number will increase in coming years because of increase in water pollution and contaminants existing in natural water. Conventional water treatment methods, such as coagulation, flocculation, sedimentation, filtration, and disinfection, have been used to remove chemicals and contaminants, which are dangerous for public health [2]. However, these methods are too slow, ineffective, waste chemicals, require large areas, are not environmentally compatible, and generate secondary harmful products. Additionally, with respect to time, there are many contaminants detected in water, which are highly toxic for human health. It was found that the existing methods are not capable of removing these hazardous materials from water and becoming more challenging with time [3,4]. Moreover, most of the countries have been formulating more and more stringent environmental laws and regulations. Nonetheless, these options will not solve water pollution problems. In view of this, there is urgent need to design and develop an advanced, cost-effective, and efficient technology that could remove a wide range of toxic chemicals and provide purified water without hampering the ecosystem.

In the past, attempts had been made to develop new and facile strategies, such as advanced oxidation processes (AOPs), membrane filtration, chemical precipitation, and UV irradiation, to overcome the drawbacks of conventional methods and provide desirable water [5,6].

Heterogeneous Nanocomposite-Photocatalysis for Water Purification.

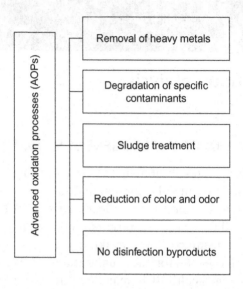

Figure 1.1 Applications of AOPs.

Among these, AOPs have gained attention because of their ability to degrade partially or eliminate totally a wide range of contaminants through oxidation and reduction reactions in water and air [7] (Figure 1.1).

AOPs are applicable in the degradation of specific pollutants, reduction of organic content, treatment of sludge, and reduction of color and odor. Additionally, heavy metals can be removed without disinfection by products [8]. Therefore, AOPs have been used in waste water treatment for the 1980s. Basically, AOPs' performance depends on the generation and utilization of hydroxyl radicals ($\cdot$OH). AOPs degrade most of the organic compounds, including carbon dioxide, because of high reduction potential (2.80 V vs. normal hydrogen electrode, NHE; acidic medium) and rate constant in the order of $10^6 - 10^9\ M^{-1}\ s^{-1}$. Various AOPs have been invented and studied in the past and can be classified mainly into four different types with respect to production of HO$^{\bullet}$ radicals (Figure 1.2) [9−11].

Among these, semiconductor-based photocatalysis or heterogeneous photocatalysis has received tremendous attention due to its potential application in wastewater treatment and production of hydrogen fuel with the help of sunlight, which is a green and abundant energy source [12]. It is defined as a "reaction assisted by photons in the presence of

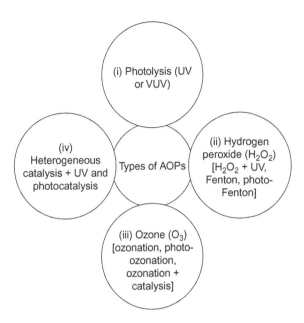

Figure 1.2 Types of AOPs.

semiconductor photocatalyst." In the beginning, the term photocatalysis was invented in different laboratories. However, as J.M. Herrmann [13] mentioned, it has been reported simultaneously by Stone and Hauffe for the adsorption/desorption and oxidation of CO on ZnO surface, respectively. At the same time, Juillet and Teichner studied the photoresponse of TiO_2 in daytime and oxidations of alkanes [14]. However, after Fujishima and Honda [15] reported water splitting into H_2 and O_2 using semiconductor photocatalysts (TiO_2), it has gained momentum in the field of science and technology. Particularly, water splitting and purification under solar light with semiconductor photocatalysts have been studied vigorously. Heterogeneous photocatalysis has attracted a lot of attention because of its enormous uses, such as degradation of organic pollutants at room temperature and pressure, almost complete mineralization of pollutant without secondary pollution, and low-cost semiconductor nanostructures. In this book, we will cover various strategies investigated on heterogeneous photocatalysis using metal oxides, chalcogenides, carbonaceous materials, and polymers. Further, we will include recent research and development progress on various aspects of photocatalysis, such as basic principle, generation and utilization of electron–hole pairs, light absorption, fabrication methods, and photocatalysis reactors used for measurements.

1.2 PRINCIPLES AND MECHANISM OF HETEROGENEOUS PHOTOCATALYSIS

The basic photochemical principles of heterogeneous photocatalysis have been reported in the literature and found that organic molecules came into contact with catalyst surface under irradiation, inducing a series of oxidation and reduction (redox) reactions and degrading pollutant molecules [16,17]. It includes degradation of toxic pollutants, organic synthesis, water splitting, metal reduction, and removal of harmful gases. Figure 1.3 shows the principle of charge transfer reaction at the semiconductor—electrolyte interface under short circuit condition. Depending on the thermodynamics of chemical reactions, photocatalysis process is divided into three classes: electrochemical photovoltaic cells, photoelectrolysis cells, and photocatalytic cells [18]. It is clear that all the photoelectrochemical cells follow a similar process of generation, separation, and transportation of electro—hole pairs after illumination of semiconductor.

After illumination, the semiconductor generates electron—hole pair, which separate due to space charge layer at the interface of semiconductor—electrolyte and, subsequently, flux of holes captured by reduced ion species of electrolyte at the interface. Afterwards, photogenerated electrons are driven to counter the electrode through an external circuit and participate in reduction reaction. Figure 1.4 shows

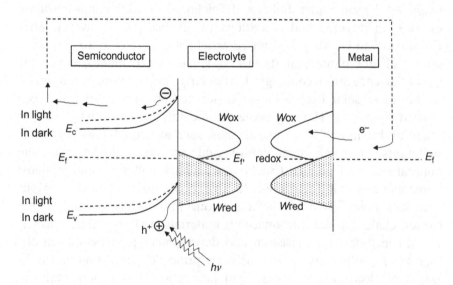

Figure 1.3 Current flow energy diagram of photoelectrochemical cell.

the types of chemical reactions that occur after irradiation of semiconductor depending on net Gibbs free energy (ΔG) gain or loss during photoelectrochemical process [19,20].

i. Electrochemical photovoltaic cell ($\Delta G = 0$): This cell converts photon to chemical energy directly. After irradiation, it generates excitons in the semiconductor particle and then electron–hole pairs are separated at the interface of semiconductor–electrolyte. The redox couple at the electrolyte remains the same for complete reaction at both anode and cathode. The electrodes are not involved in the chemical reaction and act as a carrier for charge transportation.

$$(\text{red})_{\text{solv}} + \text{h}^+ \rightleftharpoons (\text{ox})_{\text{solv}} \quad \text{at working electrode} \quad (1.1)$$

$$(\text{ox})_{\text{solv}} + \text{e}^- \rightleftharpoons (\text{red})_{\text{solv}} \quad \text{at counter electrode} \quad (1.2)$$

This cell is regenerative and could produce electricity directly under solar light.

ii. Photoelectrolysis cell ($\Delta G > 0$): These facilitate chemical energy storage of solar light by direct splitting of water into hydrogen and oxygen. Such cells are useful in direct storage of solar energy at ambient temperature and pressure. The chemical reactions in this cell are non-spontaneous and net change in ΔG is positive. The overall reaction occurring in this cell is given below:

$$H_2O \overset{h\nu}{\rightarrow} H_2 + \frac{1}{2}O_2 \quad (1.3)$$

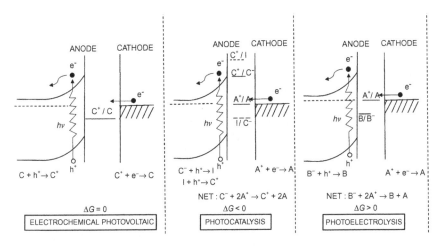

Figure 1.4 Types of photoelectrochemical process.

$$CO_2 + H_2O \rightarrow CH_2O + O_2 \qquad (1.4)$$

Therefore, this type of cell generates hydrogen fuel directly using solar light.

iii. Photocatalytic cell ($\Delta G < 0$): This cell has net negative chemical energy change during photocatalysis reactions. The incident photons only activate the chemical reaction as shown below:

$$N_2 + 3\,H_2 \xrightarrow{h\nu} 2\,NH_3 \qquad (1.5)$$

Semiconductor photocatalysis process is quite similar to the photo-electrochemical cell. Mainly, it involves three major steps: (i) light absorption and generation of electron—hole pairs; (ii) separation of charge carriers; and (iii) oxidation and reduction reactions at the surface of semiconductor. Figure 1.5 shows the basic schematic of semiconductor photocatalysis [21]. Once the semiconductor is irradiated with an energy equal or greater than the band gap electrons (e^-) are promoted from valence band (VB) to conduction band (CB) leaving holes (h^+) behind. Thereafter, excited electrons in the CB and h^+ generated in the VB migrate toward the surface of the semiconductor (i). Simultaneously, a large percentage of electrons in CB recombines with the holes in VB due to electrostatic force of interaction, leading photons,

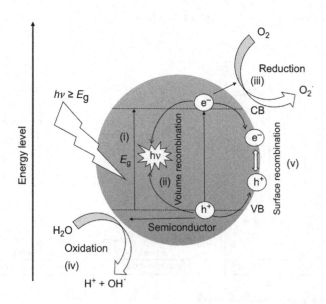

Figure 1.5 Schematic of semiconductor nanoparticle and photocatalysis mechanism.

or reproductive heat (ii). Then, CB electrons having chemical potential between +0.5 and −1.5 V vs. NHE initiates the reduction reaction and reduces acceptor species (iii). At the same time, holes migrate on the surface from VB having a chemical potential of +1.0 to +3.5 V vs. NHE oxidizing donor species adsorbed on the surface of semiconductor (iv). In addition, surface recombinations also exist due to the presence of a number of active surface states on the semiconductor crystal (v).

Usually, it is accepted that the degradation of pollutants consist of many kinds of reactions after illumination of semiconductor particles [22]. Here, we have summarized electron−hole pair generation, separation, transportation, and degradation of pollutant through oxidation−reduction reaction:

$$semiconductor + h\nu \rightarrow h^+ + e^- \qquad (1.6)$$

$$h^+ + H_2O \rightarrow \cdot OH + H^+ \qquad (1.7)$$

$$h^+ + OH^- \rightarrow \cdot OH \qquad (1.8)$$

$$h^+ + pollutant \rightarrow (pollutant)^+ \qquad (1.9)$$

$$e^- + O_2 \rightarrow \cdot O_2^- \qquad (1.10)$$

$$\cdot O_2^- + H^+ \rightarrow \cdot OOH \qquad (1.11)$$

$$\cdot OOH \rightarrow O_2 + H_2O_2 \qquad (1.12)$$

$$H_2O_2 + \cdot O_2^- \rightarrow \cdot OH + OH^- + O_2 \qquad (1.13)$$

$$H_2O_2 + \cdot O_2^- \rightarrow \cdot OH + OH^- + O_2 \qquad (1.14)$$

$$H_2O_2 + h\nu \rightarrow 2 \cdot OH \qquad (1.15)$$

$$pollutant + (\cdot OH, h^+, \cdot OOH \text{ or } O_2^-) \rightarrow pollutant \text{ degradation} \qquad (1.16)$$

From Eqs. (1.6)–(1.16), it is clear that photocatalysis process involves a series of steps, which includes generation of ˙OH radicals. These radicals have strong oxidizing power and could degrade many organic pollutants as well as converting bioresistant materials into harmless products.

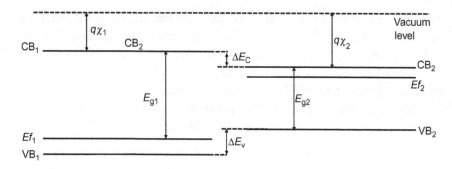

Figure 1.6 Schematic of semiconductor heterojunction.

1.3 SEMICONDUCTOR/SEMICONDUCTOR HETEROGENEOUS PHOTOCATALYSIS

To design an efficient photocatalyst, we should think about band gap energy of semiconductor and energy levels for rapid transportation of photoelectrons with minimum recombination losses. The term hetero-junction/heterostructure means junctions between two different/similar semiconductors or metal-semiconductors for devices with relatively superior performance, which is not attainable using homojunction technology [23].

The interfaces of heterojunction exhibit interesting and useful prop-erties with dramatic change in band structure at junction. Most of the existing light-harvesting devices consist of heterojunction of semicon-ductor or metals with better properties, compared to devices fabricated with homojunction. Basically, offsets of CB and VB at the interface are the crucial parameters in the formation of heterojunction. Figure 1.6 shows the band diagram of heterojunction with n/p-doped semiconductors. The relative band alignment could be explained using electron affinity (χ, energy required to move electron from the bottom of CB to vacuum level) or L. Anderson rule [24]. The electron affinity indicates the CB offset (ΔE_C) at the interface for two semiconductors (Eqs. (1.17) and (1.18)).

$$\Delta E_C = q(\chi_2 + \chi_1) \tag{1.17}$$

And VB offset is then

$$\Delta E_v = (E_{g1} - E_{g2}) - \Delta E_C = \Delta E_g - \Delta E_C \tag{1.18}$$

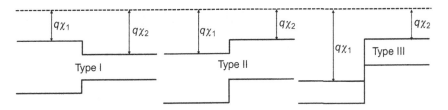

Figure 1.7 Types of semiconductor depending on interface: (I) straddling gap; (II) staggered gap; and (III) broken gap.

From the difference between electron affinities and discontinuity produced from band alignment of two semiconductors, we can classify heterojunction into three major types: Type I, Type II, and Type III (Figure 1.7) [25].

i. Type I (straddling) heterojunction: The band gap energy of one semiconductor completely overlaps with that of other semiconductor and produces discontinuity at the interface. It could be explained using Eqs (1.19) and (1.20):

$$\Delta E_C = (E_{C1} - E_{C2}) = f(\Delta E_{g1} - \Delta E_{g2}) = f \Delta E_g \qquad (1.19)$$

$$\Delta E_V = (E_{V1} - E_{V2}) = (1 - f)(E_{g1} - E_{g2}) = (1 - f)\Delta E_g \qquad (1.20)$$

ii. Type II (staggered) heterojunction: The band gap energy of the first semiconductor is lower than that of the second semiconductor, while the electrons are confined in the first and the holes in the second. Equation (1.21) shows the condition of Type II heterojunction:

$$\Delta E_C < E_{g1} \qquad (1.21)$$

iii. Type III (misaligned) heterojunction: it is formed between semimetal with inverted bands and semiconductor (Eq. (1.22)):

$$\Delta E_C > E_{g1} \qquad (1.22)$$

In different types of heterojunctions, electrons flow from high to low Fermi energy level until a potential difference at the interface becomes similar and forms space charge layer because of carrier migration. Therefore, heterojunction systems reduce charge recombination and increase the range of optical absorbance resulting in a better performance of photocatalysts. Table 1.1 summarizes the different types of heterojunction reported in the literature [26]. It was found that heterojunction improved photocatalytic efficiency.

Table 1.1 Different Types of Heterojunction

	Water Splitting	Degradation of Organic Pollutants
Type I	$CdSe/CdS$, CdS/ZnS, In_2O_3/In_2S_3	Bi_2S_3/CdS, $V_2O_5/BiVO_4$
Type II	CdS/TiO_2, $CdSe/TiO_2$, $SrTiO_3/TiO_2$, Fe_2O_3/TiO_2, ZnO/CdS, In_2O_3/Ta_2O_5, Fe_4N/Fe_2O_3, $AgIn_5S_8/TiO_2$, $TiO_2/CdS/CdSe$, $ZnO/CdS/CdSe$, $CdS\text{-}CdSSe/CdSe/TiO_2$, $ZnO/CdSSe$	CdS/TiO_2, Bi_2S_3/TiO_2, WO_3/TiO_2, Fe_2O_3/TiO_2, In_2O_3/TiO_2, $ZnFe_2O_4/TiO_2$, Fe_2O_3/WO_3, $ZnO/CdTe$, ZnO/In_2S_3, ZnO/CdS, $SrTiO_3/TiO_2$, $BiVO_4/CeO_2$, $\alpha\text{-}Fe_2O_3/CdS$, $CdS/ZnFe_2O_4$, $CdS/CoFe_2O_4$, Ag_3VO_4/TiO_2
p−n	$CuFe_2O_4/TiO_2$, CuO/ZnO, $CaFe_2O_4/TaON$, MoS_2/CdS	Ag_2O/TiO_2, $CuInSe_2/TiO_2$, TiO_2/ZnO, $ZnFe_2O_4/TiO_2$, NiO/ZnO, CuO/ZnO, Cu_2O/ZnO, $CuInS_2/ZnO$, $CuInSe_2/ZnO$, CuO/In_2O_3, $Ag_3PO_4/BiVO_4$, $Co_3O_4/BiVO_4$
Homojunction	anatase/rutile TiO_2, $\alpha/\beta\text{-}Ga_2O_3$, p−n Cu_2O, p−n Mg-doped Fe_2O_3/Fe_2O_3, W-doped $BiVO_4$, twin-induced $Cd_{0.5}Zn_{0.5}S$	N-doped anatase/brookite TiO_2, anatase/rutile TiO_2, $\alpha/\gamma\text{-}Bi_2O_3$, QDs/nanosheets Bi_2WO_6, p−n Cu_2O, p−n Co-doped TiO_2/TiO_2, p−n Fe-doped TiO_2/TiO_2, p−n Co-doped zincblende/wurtzite ZnO, $Pt/n\text{-}Si/n + - Si/Ag$

1.4 HOMOJUNCTION PHOTOCATALYSTS (SAME MATERIAL)

From the above-mentioned heterojunctions it was found that semiconductors with different band gap energy and flat band potential reduce charge recombination. It has been reported that the charge separation efficiency improved remarkably through the construction of junction within the semiconductors [27]. The gradient in the CB and VB position of same material with different size facilitates the transfer of charge carriers leads to minimum recombination. It was defined that homojunction photocatalysts are formed with same band gap energy semiconductor interfaces and chemical composition [28]. Their optical, electrical, physical, and chemical properties are strongly dependent on the phase of crystal. These kinds of junction play an important role in the fabrication of nanoscale devices, which have better performance and are technologically applicable. Various semiconductor homojunctions were reported in the literature to increase photocatalytic activity. One of the best examples is bismuth oxide (Bi_2O_3) with α and γ phases because of difference in their CB and VB positions. Sun et al. [29] successfully prepared a homojunction of $\alpha/\gamma\text{-}Bi_2O_3$ using hydrothermal method. Figure 1.8 illustrates the relative band positions as well as charge transfer and separation on the interface of $\alpha/\gamma\text{-}Bi_2O_3$ phases in which $\alpha\text{-}Bi_2O_3$ acts as a hole capture and $\gamma\text{-}Bi_2O_3$ acts as an electron sink during photocatalysis process. The obtained flat band potential values for $\alpha/\gamma\text{-}Bi_2O_3$ (-0.50 eV for $\alpha\text{-}Bi_2O_3$ and -0.45 eV for $\gamma\text{-}Bi_2O_3$)

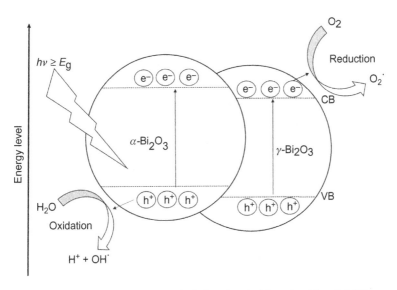

Figure 1.8 Homojunction of α/γ-Bi₂O₃ grown using hydrothermal method. Reproduced from Ref. [29].

were different leading to the formation of space charge layer at the interface and hence improving the charge separation efficiency.

Jiang et al. [30] deposited homojunction films of p–n Cu_2O using electrodeposition method and proved the high charge separation efficiency because the generation of a large amount of interfacial electric field at junction leads to better photocatalytic activity. A unique junction of homogeneous topotactic structure of BiOCl (001) showed a rapid charge separation across the interface (Figure 1.9). A layer of teethlike ultrathin nanosheets are integrated on thick nanosheets of BiOCl to obtain topotactic structure.

Weng et al. [31] achieved superior photocatalytic activity, which aroused from UV/vis light absorbance and the fast interfacial charge-transfer at homojunction. Li et al. reported Bi_2WO_6 homojunction, with quantum dots sensitized on the surface of nanosheets. Addition of quantum dots increased optical absorbance leading to the generation of a large amount of charge carriers [32]. Due to size difference, the position of bands are different resulting in high optical absorbance (Figure 1.10). After irradiation, the electron–hole pairs are generated on both nanostructures of Bi_2WO_6. A higher effective mass of electrons requires longer time to reach the same destination compared with that of holes in the same crystal.

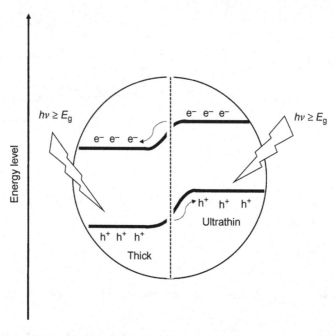

Figure 1.9 Homojunction BiOCl interfaces between thick and ultrathin nanostructures. Reproduced from Ref. [30].

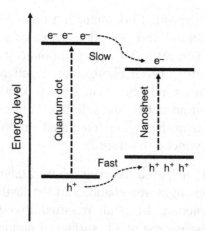

Figure 1.10 Homojunction interface between Bi_2WO_6 quantum dot and nanosheet. Reproduced from Ref. [30].

Hence, the faster velocity of holes than that of electrons leads to more number of electrons in CB of quantum dots and holes in VB of nanosheets. This process increased the separation performance of charge carriers resulting in an excellent photocatalytic activity. Therefore, it was seen that homojunction between two different thicknesses of the same materials has a synergetic effect and is a new approach for the development of efficient photocatalysts.

1.5 METAL/SEMICONDUCTOR (SCHOTTKY JUNCTION)

It is known that effective generation and transportation of charge carriers are the essential parameters for efficient photocatalysts. We mentioned about heterojunction and homojunction between semiconductor and semiconductor materials, which have shown excellent performance and a degraded wide range of contaminants. In addition, there is one more approach reported in the literature, that is, heterojunction between metal and semiconductor [23,33]. It was studied that loading of metal nanoparticles on the surface of semiconductor nanoparticles surmounted some of the limitations of photocatalysts, such as (1) reducing carrier recombination by increasing charge separation and hence efficient photocatalytic process; (2) covering response range of photocatalysts over a wide range; (3) increasing the selectivity toward degradation of particular contaminant; and (4) acting as a trapping center to enhance lifetime of carriers.

The surface-attached metal nanoparticles generate intense internal electric field close to the surface leading to effective charge separation. At the interface, electrons flow from semiconductor to metal because of lower Fermi energy level of metal and remain continues until two energy levels becomes equal. Because of charge migration from either side of junction, a potential barrier is formed. This will accumulate charge carriers, which result in the formation of a space charge layer to sustain electrical neutrality [34]. This process tends to band bending at the interface and this Schottky barrier (Φ_B) could be explained by Eq. (1.23) (Table 1.2):

$$\Phi_B = \Phi_M - X_S \qquad (1.23)$$

where Φ_M and X_S are metal work function and electron affinity of semiconductor, respectively.

As shown in Figure 1.11(a), after irradiation electron migrates to the metal nanoparticles where they are trapped. Hence, metal nanoparticles act as an electron sink for photogenerated electrons, which reduces the recombination losses. Additionally, trapping increases improved lifetime of charge carriers, which resulted in a better photocatalytic

Table 1.2 Work Function of Metal Nanoparticles	
Metal Name	Work Function (ϕ_M) in eV
Pt	6.30
Pd	5.40
Rh	4.90
Au	4.80
Ru	4.70
Cu	4.18

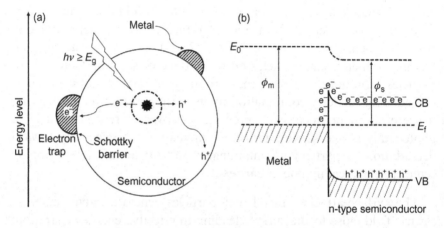

Figure 1.11 (a) Schottky barrier between n-type semiconductors and metal nanoparticles and (b) band bending of semiconductor at the interface after contact. Reproduced from Ref. [23].

performance. Table 1.3 has been reproduced with additional references, which show that most of these photocatalysts have superior or higher efficiency with metal nanoparticles compared with that of bare photocatalyst [35]. Zhang et al. [36] encapsulated TiO_2 nanospheres with Au using sol−gel and calcination method with sandwich structure with SiO_2 and proved that encapsulation increases the contact area between Au and TiO_2 matrix allowing efficient electron transfer (Figures 1.12 and 1.13).

Tian and Tatsuma reported the highly stable Au−TiO_2 system with photocatalytic deposition process. Further, they showed rapid charge separation without corrosion of Au nanoparticles (AuNPs), which reduces the cost of expensive organic/inorganic dyes [37]. Yang et al. reported synthesis of a large-scale and highly efficient photocatalyst based on single-layered unaggregated AuNPs attached to ZnO

Table 1.3 Metal Nanoparticles Sensitized Metal Oxide Heterogeneous Photocatalysts

Types	Preparation Method	Improved Photocatalytic Activity	Proposed Reasons
Au/TiO_2	Multicomponent assembly approach	Almost 3 times higher than that of TiO_2 for phenol decomposition under UV-vis irradiation	Enhanced light absorption and improved quantum efficiency
Ag-AgCl	Ion-exchange reaction combined with irradiating under visible light	8 times than N-doped TiO_2 for the decomposition of methyl orange dye in solution	Plasmon resonance and electron conductivity of Ag
Ag-AgBr	A one-step hydrothermal process	13 times and 1.5 times than that of $N-TiO_2$ and AgBr, respectively, for the degradation of RhB under visible light	Enhanced separation and transportation of photogenerated charge
$Ag-Ag_2PO_4$	A modified polyol process combined with a hetero-growth process	It can completely degrade RhB dye in only 2 min, while the pure Ag_3PO_4 cubes need about 8 min	The rapid electron export through Ag nanowires as well as highly efficient charge separation at the contact interfaces
$Ag/ AgCl@TiO_2$	One spot deposition–precipitation method	Cr(VI) removal time reduced from 30 to 5 min for $Ag/ AgCl@TiO_2$ compared to Ag/TiO_2	The coexistence of Ag and AgCl played a key role in the enhanced photocatalytic activity
Ag@C	Hydrothermal process	RhB and gaseous acetaldehyde were decomposed under visible light using Ag@C, which is much higher than that of $N-TiO_2$	Improved photocatalytic activity was ascribed to surface plasmon resonance (SPR) effect of silver nanoparticles in the Ag@C composite

nanorod arrays (ZNA) through molecular linker molecule of thioctic acid (TA) through conductive film substrate process. They have shown almost 8.1 times higher photocatalytic performance for Au-sensitized ZNA compared with that of bare ZNA toward degradation of rhodamine B under UV irradiation [38]. The excellent performance was attributed to separation of charge carriers and reaction space resulting in an effective degradation of pollutants. Under UV irradiation, generated photoelectrons in ZNA are captured by AuNPs because of suitable levels, and these electrons produce hydroxyl radical for further degradation mechanism (Figure 1.14).

In addition to the above-mentioned Schottky heterojunctions, other semiconductor nanostructures, such as g-C_3N_4, CdS, Fe_2O_3, WO_3, Ag_3PO_4, $BiVO_5$, BiOCl, SnO_2, and ZnS, have been reported in the literature [39]. These photocatalysts degraded a large number of

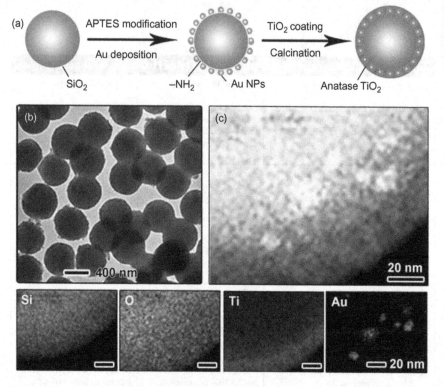

Figure 1.12 (a) Schematic illustration of the fabrication process of the sandwich-structured $SiO_2/Au/TiO_2$ photo-catalyst; (b) typical TEM image of the composite photocatalyst; and (c) elemental mapping of a single particle with the distribution of individual elements shown in the bottom row. Reproduced from Ref. [36].

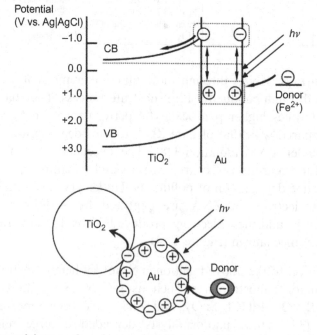

Figure 1.13 Proposed charge transfer mechanism under visible light irradiation in gold nanoparticle−TiO_2 system. Reproduced from Ref. [37].

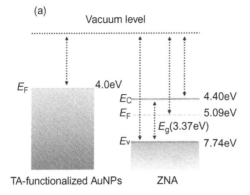

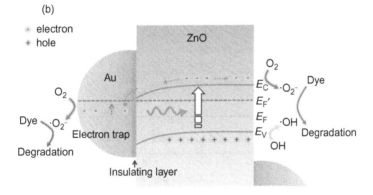

Figure 1.14 (a) Energy band diagram of isolated TA-functionalized AuNPs adjacent to the isolated ZNA under thermal nonequilibrium conditions and (b) photoinduced charge separation of the ZNA–AuNPs heterostructures. Reproduced from Ref. [38].

contaminants under visible irradiation with short time compared with that of bare nanostructure. Most of these reports have proven that metal nanoparticles with relatively high work function act as trapping sites and improve the life of charge carriers. Therefore, they have shown excellent photocatalytic activity than that for pure semiconductor nanostructures.

1.6 PHOTOCATALYTIC MATERIALS

For the development of superior heterogeneous photocatalysts, there are few basic materials requirements. Most of the semiconductors have been combined and used in the past to design efficient photocatalysts. We explained photocatalysis mechanism in Section 1.1 in which the generation of a number of charge carriers depends on the amount of light absorption by semiconductor. The first and most important is band gap energy of semiconductor. It is known that out of the total solar spectrum

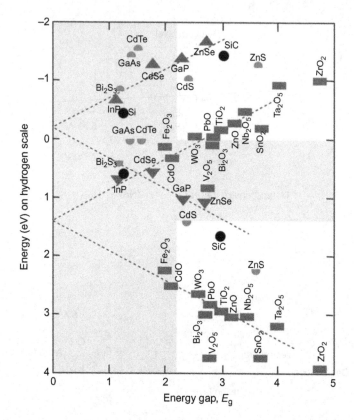

Figure 1.15 Band gap energies and band positions for some semiconductors.

around 42% exhibits visible light. This visible spectra has an energy range from 2.43 eV to 3.2 eV [40]. If we are able to capture this whole spectrum without scattering, then photocatalysts will provide desirable energy for human beings. Therefore, we need to design a semiconductor that has a band gap energy in the visible region of solar spectrum. Figure 1.15 shows the band gap energy values for different semiconductors, which have been used in photocatalysis [41]. The semiconductors having UV band gap are less suitable for photodegradation compared with that of visible band gap semiconductors. Various approaches have been investigated in the past to reduce band gap of UV-active semiconductors in order to overlap with the solar spectrum. The maximum absorption by the semiconductor can be calculated using Eq. (1.24) [25]:

$$\alpha = \frac{A(h\upsilon - E_g)^n}{h\upsilon} \tag{1.24}$$

where A is constant; $n = 1/2$ for direct band gap semiconductor; $n = 2$ for indirect band gap semiconductor; and α is absorption coefficient.

For the highest efficiency calculation, absorption coefficient should be equal to 1 (neglecting all kinds of losses): hypothetical efficiency (η_{hyp}) could be calculated using Eq. (1.25):

$$\eta_{hyp} = \frac{E_g \int_{E_g}^{\infty} N(E) \cdot dE}{\int_0^{\infty} E \cdot N(E) \cdot dE} \tag{1.25}$$

If we calculate the total hypothetical efficiency, then it would be $\approx 47\%$ at $E_g = 1.2$ eV and AM1. From this efficiency we could say that band gap of semiconductors should be between 1.4 and 1.6 eV for efficient photocatalysts [42].

The second important parameter is that the position of VB should be lower than the oxygen oxidation potential and CB should be higher than the hydrogen reduction potential. Most of the toxic chemicals are dissolved in water and water splitting will take place at 1.23 eV. Therefore, to produce a large number of photoelectrons, the band position of semiconductors should be higher than the energy required for water splitting. Moreover, the band edge of CB should be higher than the reduction potential of hydrogen and VB position should be lower than the oxidation potential of water, respectively (Figure 1.16) [43]. The third material requirement is intrinsic generation and separation of charge carriers at the interface of semiconductor junction. Once the electrons transferred on to the surface of semiconductor through diffusion process and takes part in redox reaction. The generation of carriers through diffusion process could be calculated using diffusion Eq. (1.26):

$$\eta_q = \frac{J}{e \cdot \varphi_0} \left[1 - \frac{\exp(-\alpha W)}{1 + \alpha \cdot L_P} \right] \tag{1.26}$$

where L_p is the minority carrier diffusion length, W the depletion layer width, φ_0 the photon flux, and η_q the quantum efficiency.

Equation (1.26) shows that semiconductor must fulfill the following conditions: (i) large absorption coefficient, (ii) large diffusion length, (iii) wide depletion layer, (iv) Debye length should be large with low donor concentration, and (v) maximum flat band potential.

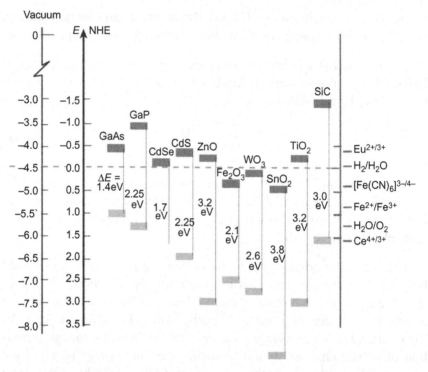

Figure 1.16 Band positions of several semiconductors in contact with aqueous electrolyte at pH 1.

The fourth material requirement is the presence of grain boundaries and surface states on the semiconductor. It is important that the generated electron–hole pairs should reach to the species of pollutant. However, due to low work function of semiconductor reversible reaction losses large quantity of carriers leads to poor photocatalytic performance. Hence, surface states existing on the surface of photocatalysts could trap the charge carriers and improve the life time. The fourth material requirement is stability against photocorrosion and dissolution. Under irradiation, photogenerated carriers could accumulate on the surface of semiconductor and pollutant, which are responsible for the photocorrosion and dissolution of photocatalysts. It was found that many semiconductors are unstable under irradiation even though they are stable in dark. To reduce this problem, it has been investigated that minimum contact between photocatalyst and pollutant could reduce the photocorrosion problem. In this approach, charge carriers could transfer through the thin protective layer. The sixth material requirement is to high specific surface for adsorption of a large quantity of pollutant molecules. At nanoscale level, materials will provide tremendous amount of sites per volume/mass to encourage the reaction. In addition to the above-mentioned parameters, few more requirements, such as electronic structure of semiconductor, geometry of grown nanostructure, potential loss at nanoscale level due to formation of deletion layer at the interface of photocatalyst, are playing a crucial role in the effective degradation of pollutants.

REFERENCES

[1] Chong MN, Jin B, Chow WK, Saint C. Recent developments in photocatalytic water treatment technology: a review. Water Res 2010;44:2997–3027.

[2] Andreozzi R, Caprio V, Insola A, Marotta R. Advanced oxidation processes (AOP) for water purification and recovery. Catal Today 1999;53:51–9.

[3] Mahamuni NN, Adewuyi YG. Advanced oxidation processes (AOPs) involving ultrasound for waste water treatment: a review with emphasis on cost estimation. Ultrason Sonochem 2010;17:990–1003.

[4] Gaya UI, Abdullah AH. Heterogeneous photocatalytic degradation of organic contaminants over titanium oxide: a review of fundamentals, progress and problems. J Photochem Photobio C Photochem Rev 2008;9:1–12.

[5] Cesaro A, Naddeo V, Belgiorno V. Wastewater treatment by combination of advanced oxidation processes and conventional biological systems. J Bioremed Biodeg 2013;4:1–8.

[6] Wang JL, Xu LJ. Advanced oxidation processes for wastewater treatment: formation of hydroxyl radical and application. Crit Rev Env Sci Tec 2012;42:251–325.

[7] Das S, Daud W. Photocatalytic CO_2 transformation into fuel: a review on advances in photocatalyst and photoreactor. Renew Sust Energy Rev 2014;39:765–805.

[8] Palmisano G, Augugliaro V, Pagliaro M, Palmisano L. Photocatalysis: a promising route for 21st century organic chemistry. Chem Commun 2007;33:3425–37.

[9] Chatterjee D, Dasgupta S. Visiblelight induced photocatalytic degradation of organic pollutants. J Photochem Photobio C Photochem Rev 2005;6:186–205.

[10] Shinde SS, Bhosale CH, Rajpure KY. Kinetic analysis of heterogeneous photocatalysis: role of hydroxyl radicals. Catal Rev Sci Eng 2013;55:79–133.

[11] Ajmal A, Majeed I, Malik RN, Idriss H, Nadeem MA. Principles and mechanisms of photocatalytic dye degradation on TiO_2 based photocatalysts: a comparative overview. RSC Adv 2014;4:37003–26.

[12] Corma A, Garcia H. Photocatalytic reduction of CO_2 for fuel production: possibilities and challenges. J Catal 2013;308:168–75.

[13] Herrmann JM. Fundamentals and misconceptions in photocatalysis. J Photochem Photobiol A Chem 2010;216:85–93.

[14] Doerffler W, Hauffe K. Heterogeneous photocatalysis II. the mechanism of the carbon monoxide oxidation at dark and illuminated zinc oxide surfaces. J Catal 1964;3:171–8.

[15] Fujishima A, Honda K. Electrochemical photolysis of water at a semiconductor electrode. Nature 1972;238:37–8.

[16] Rauf MA, Ashraf SS. Fundamental principles and application of heterogeneous photocatalytic degradation of dyes in solution. Chem Eng J 2009;151:10–18.

[17] Ohtani B. Revisiting the fundamental physical chemistry in heterogeneous photocatalysis: its thermodynamics and kinetics. Phys Chem Chem Phys 2014;16:1788−97.

[18] Linsebigler AL, Lu G, Yates JT. Photocatalysis on TiO_2 surfaces: principles, mechanisms, and selected results. Chem Rev 1995;95:735−58.

[19] Kudo A. Photocatalyst materials for water splitting. Catal Surveys Asia 2003;7:31−8.

[20] Chen X, Shen S, Guo L, Mao SS. Semiconductor-based photocatalytic hydrogen generation. Chem Rev 2010;110:6503−70.

[21] Fox MA, Dulay MT. Heterogeneous photocatalysis. Chem Rev 1993;93:341−57.

[22] Kisch H. Semiconductor photocatalysis-mechanistic and synthetic aspects. Angew Chem Int Ed 2013;52:812−47.

[23] Wang H, Zhang L, Chen Z, Hu J, Li S, Wang Z, et al. Semiconductor heterojunction photocatalysts: design, construction, and photocatalytic performances. Chem Soc Rev 2014;43:5234−44.

[24] Miles AG. Semiconductor heterojunction topics: introduction and overview. Sol State Ele 1986;29:99−121.

[25] Chandra S. Charge transfer reactions at the semiconductor-electrolyte interface. Photoelectrochemical Solar Cells, 5. New York: Gordon and Breach; 1985.

[26] Li H, Zhou Y, Tu W, Ye J, Zou Z. State-of-the-art progress in diverse heterostructured photocatalysts toward promoting photocatalytic performance. Adv Funct Mater 2015;25:1−16.

[27] Huang ZF, Pan L, Zou JJ, Zhang X, Wang L. Nanostructured bismuth vanadate-based materials for solar-energy-driven water oxidation: a review on recent progress. Nanoscale 2014;6:14044−63.

[28] Feng X, Hu G, Hu J. Solution-phase synthesis of metal and/or semiconductor homojunction/heterojunction nanomaterials. Nanoscale 2011;3:2099−117.

[29] Sun Y, Wang W, Zhang L, Zhang Z. Design and controllable synthesis of α-/γ-Bi_2O_3 homojunction with synergetic effect on photocatalytic activity. Chem Eng J 2012;211−212:161−7.

[30] Jiang T, Xie T, Yang W, Chen L, Fan H, Wang D. Photoelectrochemical and photovoltaic properties of p−n Cu_2O homojunction films and their photocatalytic performance. J Phys Chem C 2013;117:4619−24.

[31] Weng S, Fang Z, Wang Z, Zheng Z, Feng W, Liu P. Construction of teethlike homojunction BiOCl (001) nanosheets by selective etching and its high photocatalytic activity. ACS Appl Mater Interfaces 2014;6:18423−8.

[32] Li C, Chen G, Sun J, Dong H, Wang Y, Lv C. Construction of Bi_2WO_6 homojunction via QDs self-decoration and its improved separation efficiency of charge carriers and photocatalytic ability. Appl Catal B Environ 2014;160−161:383−9.

[33] Khan MR, Chuan TW, Yousuf A, Chowdhury MNK, Cheng CK. Schottky barrier and surface plasmonic resonance phenomena towards the photocatalytic reaction: study of their mechanisms to enhance photocatalytic activity. Catal Sci Technol 2015. Available from: http://dx.doi.org/10.1039/C4CY01545B.

[34] Hou W, Cronin SB. A review of surface plasmon resonance-enhanced photocatalysis. Adv Funct Mater 2013;23:1612−19.

[35] Wang Z, Liu Y, Huang B, Dai Y, Lou Z, Wang G, et al. Progress on extending the light absorption spectra of photocatalysts. Phys Chem Chem Phys 2014;16:2758−74.

[36] Zhang Q, Lima DQ, Lee LI, Zaera F, Chi M, Yin Y. A highly active titanium dioxide based visible-light photocatalyst with nonmetal doping and plasmonic metal decoration. Angew Chem 2011;123:7226−30.

[37] Tian Y, Tatsuma T. Mechanisms and applications of plasmon-induced charge separation at TiO_2 films loaded with gold nanoparticles. J Am Chem Soc 2005;127:7632−7.

[38] Yang TH, Huang LD, Harn YW, Lin CC, Chang JK, Wu CI, et al. High density unaggregated Au nanoparticles on ZnO nanorod arrays function as efficient and recyclable photocatalysts for environmental purification. Small 2013;9:3169−82.

[39] Xiao FX, Miao J, Tao HB, Hung SF, Wang HY, Yang HB, et al. One-dimensional hybrid nanostructures for heterogeneous photocatalysis and photoelectrocatalysis. Small 2015. Available from: http://dx.doi.org/10.1002/smll.201402420.

[40] Guo JF, Ma B, Yin A, Fan K, Dai WL. Highly stable and efficient Ag/AgCl@TiO_2 photocatalyst: preparation, characterization, and application in the treatment of aqueous hazardous pollutants. J Hazard Mater 2012;211−212:77−82.

[41] Li L, Salvador PA, Rohrer GS. Photocatalysts with internal electric fields. Nanoscale 2014;6:24−42.

[42] Wei D, Andrew P, Ryhanen T. Electrochemical photovoltaic cells—review of recent developments. J Chem Technol Biotechnol 2010;85:1547−52.

[43] Gratzel M. Photoelectrochemical cells. Nature 2001;414:338−44.

CHAPTER 2

Nanomaterial-Based Photocatalysis

2.1 INTRODUCTION

Since the last few decades, nanostructured materials (NsM) have been explored and investigated curiously worldwide in various applications including energy and environmental. NsM could be defined as the solids composed of structural elements—mostly crystallites—with a characteristic size (in at least one direction) of a few nanometers (1–100 nm). These materials exhibit outstanding and often superior physical and chemical properties compared with their bulk counterpart because of different chemical composition, arrangement of the atoms, and size. The growth of NsM in one, two, and three dimensions generates new interesting properties, which remarkably enhanced functions for device fabrication in various fields, including energy, medical, biological, opto-electronics, optics, magnetic, electronic, and many others. Particularly, metal and metal oxide nanostructures have been studied potentially due to their high specific surface to volume ratio, showed quantum size effect, properties can be controlled just adjusting shape and size and high interfacial reactivity. NsM could be classified into four types—zero: clusters of any aspect ratio from 1 to ∞; one: multilayers; two: ultrafine-grained overlayers or buried layers; and three: nanophase materials depend on their dimensional grain growth (Figure 2.1). The properties of NsM are examined by their size distribution, shape, chemical composition and interfacial reactivity, and type of grains present at the interfaces [1]. The change in the size makes these materials have different electronic changes in terms of energy and number of levels. This makes these materials behave electronically different. The origin of the size-induced properties in nanomaterials depends basically on the surface phenomena (extrinsic contribution) and quantum confinement effects (intrinsic contribution). This chapter gives a detailed summary about NsM and their outstanding properties [2].

Heterogeneous Nanocomposite-Photocatalysis for Water Purification.

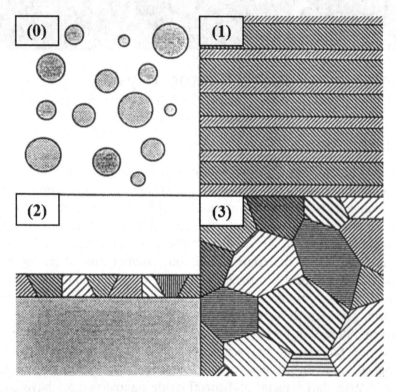

Figure 2.1 Schematic of the four types of nanostructured materials. Reproduced from Ref. 1.

2.2 PROPERTIES OF NsM

2.2.1 Increase in Surface Area to Volume Ratio

First, nanomaterials have a relatively larger surface area compared with the same volume (or mass) of the material produces in a larger form. Let us consider a sphere of radius "r," surface area, and volume of sphere given in Eqs. (2.1) and (2.2). The ratio of these two equations gives Eq. (2.3), which indicates the increase in surface to volume ratio with decrease in size of materials. Thus, when the radius of the sphere decreases, its surface area to volume ratio increases [3,4]. Figure 2.2 shows the relation between surface area and size of silica nanoparticles. It is seen that obtained surface area increased exponentially below 100 nm indicated crucial role of size in NsM. If a bulk material is subdivided into an ensemble of individual nanomaterials, the total volume remains the same, but the collective surface area is greatly increased.

$$\text{Its surface area } A = 4\pi r^2 \tag{2.1}$$

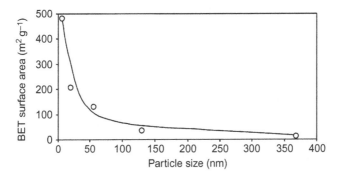

Figure 2.2 Effect of size on surface area of silica nanoparticles.

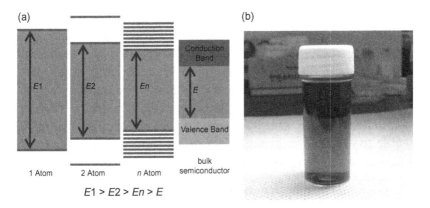

Figure 2.3 (a) Dependence of a quantum-sized semiconductor's band gap on particle size and (b) gold color in the nano form.

$$\text{Its volume } V = \left(\frac{4}{3}\right)\pi r^3 \qquad (2.2)$$

$$\text{Surface area to its volume ratio } \frac{A}{V} = \frac{3}{r} \qquad (2.3)$$

2.2.2 Quantum Confinement Effect

Contemporary literature on ultradisperse semiconductors distinguishes between the effects arising as a result of increase in surface area and the degree of surface imperfection with decrease in the size of the crystals and, separately, quantum size effects due to radical change in the electronic state of the semiconductor (Figure 2.3a) crystals less than a certain "critical" size, determined in turn by the extent to which the electron–hole pair photogenerated in the semiconductor is delocalized [5]. These effects also differ in the range of sizes in which they appear. Size effects are observed in semiconductor crystals measuring 10–100 nm, whereas quantum size effects are usually characteristic of nanocrystallites measuring less than 10 nm. In the literature semiconductor nanoparticles in which quantum size effects of one type or another appear, are often called quantum size particles or quantum points to emphasize their special electronic structure. It is necessary to mention the tentative nature of the classification in so far as the exact "critical" size after which the appearance of quantum size effects can be expected is largely determined by the chemical nature of the semiconductor and can vary from 0.5 (CuCl) to 46 (PbSe) or more nanometers. From the positions of quantum mechanics the "critical" size (the threshold for the appearance of quantum size effects) corresponds to the De Broglie wavelength of the free electron. During analysis of the interband absorption of the semiconductor nanoparticle the Bohr radius of the exciton (a_B), which can be calculated from the electrophysical constants of the bulk semiconductor, can be used as such a criterion [6]. The data that have accumulated on size effects in semiconductor nanoparticles make it possible to examine them according to the nature of the effect on the properties of the nanocrystals.

Restriction of the free motion of the exciton in the bulk of the nanocrystal leads to an increase of its energy (E_g^{nano}) in relation to the volume (bulk) semiconductor (E_g^{bulk}). The increase of the energy of the exciton as a result of the quantum size effect ($\Delta E = E_g^{nano} - E_g^{bulk}$) can be calculated in the approximation of effective masses, which is based on assumptions about the parabolic nature of the permitted energy bands close to their edges and the invariability of the effective masses of the electron of the conduction band (m_e^*) and the hole of the valence band (m_h^*) in the transition from the bulk to the ultradisperse semiconductors. This approximation gives the following expression for ΔE [7],

$$\Delta E = \frac{\pi^2 \hbar^2}{2R^2} \left(\frac{1}{m_e^*} + \frac{1}{m_h^*} \right) - \frac{1.786 e^2}{\varepsilon R} - 0.248 R_y^* \qquad (2.4)$$

The first term in Eq. (2.4) depends on R^2 and corresponds to the increase in the energy of the exciton as a result of its spatial restriction in the potential box—the semiconductor nanocrystal. The second term determines the energy of columbic interaction of the electron and the hole in the composition of the exciton and increases with decrease in the size of the nanoparticle. The value of R_y^* in the third term is called the Rydberg energy of the exciton and takes account of the correlation between the motion of the electron and the hole. As seen, the last two terms in Eq. (2.4) lead to a decrease in the energy of the exciton, which is restricted in the volume of the particle. Conventionally, it is known that the color of gold is golden, but at the nanoscale it starts to change dramatically due to quantum size effect. It is found that the colloidal solution of gold nanoparticles is no longer golden but ruby-red in color (Figure 2.3b).

A thin film of gold deposit absorbs across most of the visible part of the electromagnetic spectrum and very strongly in the IR and at all longer wavelengths. It dips slightly around 400–500 nm, and when held up to the light, such a thin film appears blue due to the weak transmission of light in this wavelength range [8]. However, the dilute gold colloid film displays total transparency at low photon energies (below 1.8 eV). Its absorption becomes intense in a sharp band around 2.3 eV (520 nm). This kind of effect is known as surface plasmon, which is aroused from small size of metal particles [9]. Besides quantum size effects, the NsM behavior is different due to surface effects, which dominate as nanocrystal size decreases. Reducing the size of a crystal from 30 to 3 nm, the number of atoms on its surface increases from 5% to 50% beginning to perturb the periodicity of the "infinite" lattice. In that sense, atoms at the surface have fewer direct neighbors than atoms in the bulk and as a result they are less stabilized than bulk atoms. The origin of the quantum size effects strongly depends on the type of bonding in the crystal [10].

In the NsM at least, one dimension being reduced to nanoscale the electronic wavevectors becomes quantized and the system exhibits discrete energy levels. The relevant length scale is the de Broglie wavelength, λ_{dB}, of an electron in one particular direction: If the dimension of the system in one dimension is lower than λ_{dB} we call it a two-dimensional (2D) system. Graphene sheet obtained from exfoliation of graphite is the best example of a 2D system as well as semiconductor superlattices. One-dimensional (1D) system is obtained when two dimensions are lower than λ_{dB} values; nanowires and nanotubes belong to this category [11]. Special is the case of the so-called quantum dots whose all three dimensions are lower than λ_{dB} and we name these as 0D systems. Their energy spectra are discrete and the system can be viewed as an artificial atom. This has a tremendous effect on optical properties of nanoparticles, as the absorption shifts from the

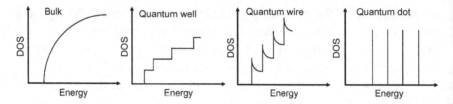

Figure 2.4 Density of states DOS for bulk semiconductor and quantum wells, quantum wires, and quantum dots, respectively.

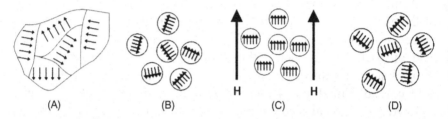

Figure 2.5 Explanation of the superparamagnetic effect.

infrared to the visible range. Quantum confinement in quantum wells (1D), quantum wires (2D), and quantum dots (3D) and the resulting changes of the energy spectrum as shown in Figure 2.4 has also serious implications for the optoelectronic properties of nanostructured semiconductors, and is currently utilized in optoelectronic devices.

2.2.3 Magnetic Effects

Ferromagnetic materials exhibit domains with parallel magnetization (Figure 2.5a). If a magnetic field H is applied, the magnetization of all domains takes the direction of the field and remains in this direction even if the outside field is removed. If the size of ferromagnetic nanoparticles becomes smaller than the critical domain size (10−20 nm), only one domain remains in the particle (Figure 2.5b). If again a magnetic field is applied, all particles will align according to this field (Figure 2.5c), but if the field is removed, thermal motion will lead to a loss of orientation (Figure 2.5d). This behavior is similar to that of permanent magnetic dipoles in a paramagnet and is thus called superparamagnetism [12]. This effect sets, among others, an upper limit to the miniaturization of magnetic memories. On the other hand, superparamagnetic particles are envisioned to play an important role in nanobiotechnology and medicine. Another magnetic nanoeffect, which is used presently in magnetic memories, is the so-called giant magnetoresistive effect [13]. Depending on the details of the realization, the critical length is either the electron mean free path or the spin relaxation length.

2.3 IMPROVED PERFORMANCE WITH NANOSTRUCTURED PHOTOCATALYSTS

Various facile and cost-effective routes have been investigated and used to design, produce, and characterize numerous NsM, including nanoparticles, nanocubes, nanorods, nanowires, and nanotubes, which maintain fundamentally interesting size-dependent chemical, physical, optical, and many other properties. From the perspective of applications, these structures have wide-ranging utility in areas as diverse as catalysis, energy storage, fiber engineering, fuel cells, biomedicine, computation, power generation, photonics, pollution remediation, and gas sensing. There are few basic parameters, such as synthesis cost, use of nontoxic chemicals, maximum use of aqueous solvents, minimum number of reagents and few steps with high yield, room or low temperature and high efficiency, which needs to be considered during fabrication of advanced devices based on NsM. Nowadays synthesis of new NsM using chemical methods have focused greatly because of low temperature and cost with better functions for environmental and energy applications. The synthesis of NsM is an interesting research field. Until now, a large number of approaches have been explored to synthesize NsM, which could be divided into two major sections, that is, top-down and bottom-up. Interestingly, bottom-up route based on chemistry have attracted considerable attention because of relatively low cost and high yield [14]. In this process, material growth started from atom to atom, molecule by molecule. Recently, a number of techniques, including coprecipitation, sol−gel processes, microemulsions, freeze drying, hydrothermal processes, laser pyrolysis, ultrasound and microwave irradiation, templates, and chemical vapor deposition, have been developed to control the growth, size, morphology, and uniformity of NsM [15]. Therefore, in this section, we will study in brief about heterogeneous photocatalysts fabricated using selective bottom-up approaches for improved photocatalytic activity.

(i) Shortened carrier collection pathways: Photoexcitation produces charge carriers with finite mobility and lifetime, depending on the material, the carrier type, and the light intensity. To drive water redox reactions, these carriers need to reach the material interfaces at the

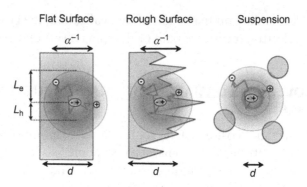

Figure 2.6 Charge collection in flat and nanostructured films and in particle suspensions: d, *film or particle thickness;* L_e, *electron diffusion length;* L_h, *hole diffusion length.*

electrolyte and at the back contact. In the absence of an external field, charge carriers move by diffusion and their range is defined by the mean free diffusion length L. The parameter L depends on the carrier diffusion constant D and the carrier lifetime τ (Eq. (2.5)), and a dimensionality factor ($q = 2, 4, 6$ for 1, 2, or 3D diffusion).

$$\overline{L}^2 = qD\tau \qquad (2.5)$$

For intrinsic semiconductors, usually $L_e > L_h$ because of the larger diffusion constant D of the electrons compared to holes. Upon doping NsM, the concentration of the majority carriers increases, and with it their τ and L values. On the contrary, the lifetime and diffusion length of the minority carriers decrease [16]. For optimum collection of both carrier types at the back contact, the semiconductor film thickness d has to be in the same range as L_e and L_h (Figure 2.6). To improve minority carrier collection at the semiconductor–electrolyte interface, the surface roughness of the film can be increased. This surface nanostructuring approach is particularly useful for first row transition metal oxides (MnO_2, Fe_2O_3), which suffer from low hole mobility and lifetimes [17]. Ideal electron/hole collection is possible with suspended nanoparticles, if their particle size $d < L_e, L_p$. Although in this situation, both carrier types need to be extracted at the sc–electrolyte interface. Thus, there is a need for selective redox agents. This has been referred to by Helmut Tributsch as kinetic rectification [18].

(ii) Improved light distribution: The ability of a material to absorb light is determined by the Lambert–Beer law and the wavelength-dependent absorption coefficient α. The light penetration depth α^{-1}

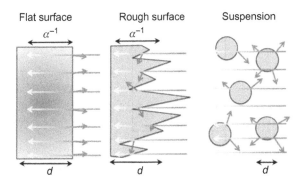

Figure 2.7 Light distribution in flat and nanostructured films, and in particle suspensions: d, *film or particle thickness.*

refers to the distance after which the light intensity is reduced to $1/e$ of the original value. For example, for Fe_2O_3, $\alpha^{-1} = 118\,nm$ at $\lambda = 550\,nm$, for CdTe, $\alpha^{-1} = 106\,nm$ (550 nm), and for Si, $\alpha^{-1} = 680\,nm$ (510 nm) [19]. To ensure >90% absorption of the incident light, the film thickness must be >2.3 times the value of α^{-1} (Figure 2.7). Surface-structuring on the micro- or nanoscale can increase the degree of horizontal light distribution by light scattering. This "trapped" light would otherwise be lost by direct reflection from a flat surface [20]. Light scattering is maximal in particle suspensions, because it can occur at both the front and back sides of the particles.

(iii) Surface area-enhanced charge transfer: The larger specific surface area of nanomaterials promotes charge transfer across the material interfaces (solid–solid and solid–liquid), allowing water redox reactions to occur at relatively low current densities and, correspondingly, low over potentials. In other words, the increase of surface area allows to better match the photocurrents with the slow kinetics of the water redox reactions [21]. In particular proceed at Fe_2O_3 and TiO_2, according to recent transient absorption measurements. Thus, increases of surface area reduce the need for highly active, and often expensive, cocatalysts, based on Ir, Rh, or Pt [22].

(iv) Multiple exciton generation: The altered electronic structure of strongly size-confined nanocrystals gives rise to multiple exciton generation (MEG), that is, the formation of several (n) electron–hole pairs after absorption of one photon with an energy n times the band gap of the dot (Figure 2.8). The MEG effect is responsible for the abnormally high efficiency of PbSe QD-sensitized TiO_2 PEC cells, and PbSe

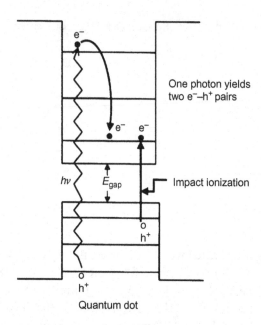

Figure 2.8 Enhanced photovoltaic efficiency in QD solar cells by impact ionization.

photovoltaic cells [23]. The effect has not yet been applied to water photoelectrolysis. Future MEG-enhanced water splitting devices will likely be Tandem or multi-junction devices, because the individual quantum dots cannot produce a sufficient potential for overall water splitting. This is because for efficient solar energy conversion, the band gaps of the relevant dots need to be a fraction of the energy of visible light photons ($E = 1.55-3.1$ eV).

2.4 APPLICATIONS OF NANOSTRUCTURED PHOTOCATALYSTS

Both the technological and economic importance of photocatalysis have increased considerably over the past decade. Improvements in performance have been strongly correlated to advances in nanotechnology; for example, the introduction of nanoparticulate photocatalysts has tremendously enhanced the catalytic efficiency of specific materials. A variety of applications ranging from anti-fogging, anti-microbial, and self-cleaning surfaces, through to water and air purification and solar-induced hydrogen production, have been developed and many of these have made their way into commercial products. However, extensive research continues to further optimize this

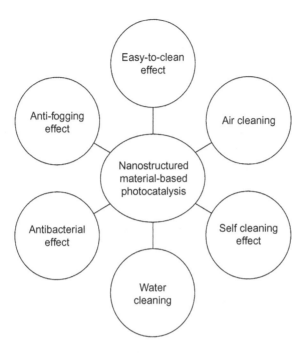

Figure 2.9 Photocatalysis applications.

technology and to widen the spectrum of potential applications. Research and application foci include anti-stick or anti-fingerprint coatings, soil repellency, and decomposition of organic matter, such as microbes or fat. When exposed to light certain semiconducting materials such as "photocatalysts" trigger or accelerate chemical reactions resulting, for example, in a decomposition of organic molecules [24]. Due to their large surface area, nanosized catalyst particles show a significantly enhanced reactivity compared to larger particles or bulk material. Numerous materials are under examination; however, none appear to match the efficiency of TiO_2. Its application requires illumination in the UV or at the extreme blue edge of the visible spectrum. Volume applications are thus mainly limited to the outdoor area [25]. However, despite the reduced natural illumination, even indoor products such as sanitary ceramics are being increasingly applied. Moreover, research is underway to widen the exploitable spectral range toward visible light. A few applications of photocatalysis are provided in Figure 2.9. In recent years, applications have been directed toward environmental clean-up, drinking water treatment, industrial, and health applications.

(i) Removing trace metals: Trace metals, such as mercury (Hg), chromium (Cr), lead (Pb), and other metals, are considered to be highly hazardous to health. Thus, removing these toxic metals is essentially important for human health and water quality. The environmental applications of heterogeneous photocatalysis include removing heavy metals such as mercury (Hg), chromium (Cr), cadmium (Cd), lead (Pb), arsenic (As), nickel (Ni), and copper (Cu). The photoreducing ability of photocatalysis has been used to recover expensive metals from industrial effluent, such as gold, platinum, and silver [26].

(ii) Destruction of organics: Photocatalysis has been used for the destruction of organic compounds, such as alcohols, carboxylic acids, phenolic derivatives, or chlorinated aromatics, into harmless products, for example, carbon dioxide, water, and simple mineral acids. Water contaminated by oil can be treated efficiently by photocatalytic reaction. Herbicides and pesticides that may contaminate water, such as 2,4,5-trichlorophenoxyacetic acid, 2,4,5-trichlorophenol, and s-triazine herbicides, can be also mineralized [27].

(iii) Removing inorganic compounds: In addition to organic compounds, wide ranges of inorganic compounds are sensitive to photochemical transformation on the catalyst surfaces. Inorganic species such as bromate, or chlorate, azide, halide ions, nitric oxide, palladium and rhodium species, and sulfur species can be decomposed [28]. Metal salts such as $AgNO_3$, $HgCl$, and organometallic compounds (e.g., CH_3HgCl) can be removed from water, as well as cyanide, thiocyanate, ammonia, nitrates, and nitrites.

(iv) Water disinfections: Photocatalysis can also be used to destroy bacteria and viruses. *Streptococcus mutans, Streptococcus natuss, Streptococcus cricetus, Escherichia coli, Saccharomyces cerevisiae, Lactobacillus acidophilus*, and poliovirus 1 were destructed effectively using heterogeneous photocatalysis [29]. The increasing incidence of algal blooms in fresh water supplies and the consequent possibility of cyanobacterial microcystin contamination of potable water microcystin toxins is also degraded on immobilized titanium dioxide catalyst. Photodisinfection sensitized by TiO_2 had some effect on the degradation of *Chlorella vulgaris* (Green algae), which has a thick cell wall.

(v) Degradation of natural organic matter: Humic substances (HS) are ubiquitous and defined as a category of naturally occurring biogenic heterogeneous organic substances that can be generally characterized as being yellow-brown and having high molecular weights [30]. These are also defined as the fraction of filtered water that adsorb on XAD-8 resin (nonionic polymeric adsorbent) at pH 2. They are the main constituents of the dissolved organic carbon pool in surface waters (freshwaters and marine waters), and ground waters, commonly imparting a yellowish-brown color to the water system. The concentration of HS varies from place to place; the values in seawater being normally from 2 to 3 mg L^{-1}. Their size, molecular weight, elemental composition, structure, and number and position of functional groups vary, depending on the origin and age of the material. HS are known to affect the behavior of some pollutants significantly in natural environments, such as trace metal speciation and toxicity, solubilization and adsorption of hydrophobic pollutants, and aqueous photochemistry. HS act as substrates for bacterial growth, inhibit the bacterial degradation of impurities (some color) in natural water, and form complex with heavy metals such as Fe, Pb, and Mn making it harder to remove them, transport the metals in the environment, and also promote the corrosion of pipes. HSs act as a source of methyl groups and thus react with hypochlorite ion, which is used as a biocide in water treatment plants, to produce disinfectant byproducts, for example, trihalomethanes, haloacetic acids, other chlorinated compounds, and nitriles, some of which are suspected to be carcinogenic. More than 150 products have been identified when HS react with chlorine [31]. Recently, 3-chloro-4-(dichloromethyl)-5-hydroxy-2(5H)-furanone [abbreviated as (MX)] was found in the chlorinated water containing HS. Advanced oxidation has been applied to decreasing the organic content in water including humic acid. It has the advantage of not leaving any toxic byproducts or sludge [32]. Heterogeneous photocatalysis was also coupled with other physical methods to increase the degradation rate of organic molecules including HA (sono-photocatalysis, ozonation photocatalysis).

(vi) Seawater treatment: Recently, HS was also decomposed in highly saline water (artificial seawater) and natural seawater using different photocatalytic materials. The decomposition rate of HS in seawater was slow compared with pure water media. No toxic byproducts were detected during the decomposition. Minero et al. [33] studied the decomposition of some components in crude oil (dodecane and toluene) in the seawater media. They found that no chlorinated compounds have been detected during the irradiation, and complete decomposition was achieved after a few hours of irradiation. Another study conducted on the decomposition of seawater-soluble crude oil fractions found that it can be decomposed under illumination of nanoparticles of TiO_2 using artificial light.

(vii) Air cleaning: The photocatalytic process is well recognized for the removal of organic pollutants in the gaseous phase, such as volatile organic compounds, having great potential applications to contaminant control in indoor environments, such as residences, office buildings, factories, aircraft, and spacecraft. To increase the scope of the photocatalytic process in the application to indoor air, the disinfection capabilities of this technique are under investigation. Disinfection is of importance in indoor air applications because of the risk of exposure to harmful airborne contaminants. Bioaerosols are a major contributor to indoor air pollution, and more than 60 bacteria, viruses, and fungi have been documented as infectious airborne pathogens. Diseases transmitted through bioaerosols include tuberculosis, legionaries, influenza, colds, mumps, measles, rubella, small pox, aspergillosis, pneumonia, meningitis, diphtheria, and scarlet fever [34]. Traditional technologies to clean indoor air include the use of activated charcoal filters, HEPA filters, ozonation, air ionization, and bioguard filters. None of these technologies is completely effective.

Photocatalytic oxidation can also inactivate infectious microorganisms, which can be airborne bioterrorism weapons, such as *Bacillus anthracis* (anthrax). A photocatalytic system was investigated by Knight in 2003 to reduce the spread of severe acute respiratory syndrome on flights, following the outbreak of the disease. Similarly, in 2007 the avian influenza virus A/H5N2 was shown to be inactivated from the gaseous phase using a photocatalytic prototype system [35]. Inactivation of various Gram-positive and Gram-negative bacteria using visible light and a doped catalyst and fluorescent light irradiation similar to that used in indoor environments was studied and shows great promise for widespread applications. It was also shown that *E. coli* could be completely mineralized on a coated surface in air. Carbon mass balance and kinetic data for complete oxidation of *E. coli*, *Aspergillus niger*, *Micrococcus luteus*, and *Bacillus subtillus* cells and spores were subsequently presented. A comprehensive mechanism and detailed description of the photokilling of *E. coli* on coated surfaces in air has been extensively studied in order to understand to a considerable degree and in a quantitative way the kinetics of *E. coli* immobilization and abatement by photocatalysis, using FTIR, AFM, and CFU as a function of time and peroxidation of the membrane cell walls. Novel photoreactors and photo-assisted catalytic systems for air disinfection applications such as those using polyester supports for the catalyst, carbon nanotubes, combination with other disinfection systems, membrane systems, use of silver bactericidal agents in cotton textiles for the abatement of *E. coli* in air, high surface area CuO catalysts, and structure silica surfaces have also been reported. In terms of environmental health, the antifungal capability of photocatalysis against mold fungi on coated wood boards used in buildings was confirmed using *A. niger* as a test microbe, and UVA irradiation [36].

2.5 CONCLUSION

The photocatalytic technique is a versatile and efficient disinfection process capable of inactivating a wide range of harmful microorganisms in various media. It is a safe, nontoxic, and relatively inexpensive disinfection method whose adaptability allows it to be used for many purposes. Research in the field of photocatalytic disinfection is very diverse, covering a broad range of applications. Particularly, the use of photocatalysis based on NsM was shown to be effective for various air-cleaning applications to inactivate harmful airborne microbial pathogens, or to combat airborne bioterror threats, such as anthrax. Photocatalytic thin films on various substrates were also shown to have a potential application for "self-disinfecting" surfaces and materials, which can be used for medical implants, "self-disinfecting" surgical tools, and surfaces in laboratory and hospital settings, and equipment in the pharmaceutical and food industries. Photocatalytic food packaging was shown to be a potential way to reduce the risk of foodborne illnesses in cut lettuce and other packaged foods. In terms of plant protection, photocatalysis is being investigated for use in hydroponic agricultures as an alternative to harsh pesticides. For water treatment applications, photocatalytic disinfection has been studied and implemented for drinking water production using novel reactors and solar irradiation. Eutrophic waters containing algal blooms were also shown to be effectively treated using coated hollow beads and solar irradiation. The effectiveness of photocatalytic disinfection for inactivating microorganisms of concern for each of these applications was presented, highlighting key studies and research efforts conducted. While the performance of this technology is still to be optimized for the specific applications, based on the literature presented, it is abundantly evident that photocatalysis should be considered a viable alternative to traditional disinfection methods in some cases.

In a move toward a more environmentally friendly world, traditional solutions to classic problems, such as the production of safe drinking water, must shift toward more sustainable alternatives. Photocatalytic disinfection is not only a replacement technology for traditional methods in traditional applications, but also a novel approach for solving other disinfection problems, such as the control of bioterror threats. In this sense, the strength of photocatalytic disinfection lies in its versatility for use in many different applications.

REFERENCES

[1] Siegel RW. Nanostructured materials—mind over matter. Nanostruct Mater 1993;3:1−18.

[2] Hu X, Li G, Yu JC. Design, fabrication, and modification of nanostructured semiconductor materials for environmental and energy applications. Langmuir 2010;26:3031−9.

[3] Gleiter H. Nanostructured materials: basic concepts and microstructure. Acta Mater 2000;48:1−29.

[4] Koch CC. Synthesis of nanostructured materials by mechanical milling: problems and opportunities. Nanostruct Mater 1997;9:13−22.

[5] Obare SO, Meyer GJ. Nanostructured materials for environmental remediation of organic contaminants in water. J Environ Sci Health Part A 2004;A39:2549−82.

[6] Alivisatos AP. Semiconductor clusters, nanocrystals, and quantum dots. Science 1996;217:933−7.

[7] Brus LE. Electron-electron and electron-hole interactions in small semiconductor crystallites— the size dependence of the lowest excited electronic state. J Chem Phys 1984;80:4403−9.

[8] Moriarty P. Nanostructured materials. Rep Prog Phys 2001;64:297−381.

[9] Oelhafen P, Schuler A. Nanostructured materials for solar energy conversion. Solar Energy 2005;79:110−21.

[10] Logothetidis S. Nanostructured materials and their applications. Springer; 2011. p. 6.

[11] Davies AG, Thompson JMT. Advances in nanoengineering electronics, materials and assembly. Royal Society Series on Advances in Science, vol. 3. London: Imperial College Press; 2007. p. 3.

[12] Reithmaier JP, Petkov P, Kulisch W, Popov C. Nanostructured materials for advanced technological applications. Springer; 2009. p. 6.

[13] Anderson JC, Leaver KD, Rawlings RD, Alexander JM. Materials Science, London: Chapman & Hall; 1990.

[14] Gleiter H. Materials with ultrafine microstructures: petrospectives and perspectives. Nanoscruct Mater 1992;1:1−19.

[15] Osterloh FE. Inorganic nanostructures for photoelectrochemical and photocatalytic water splitting. Chem Soc Rev 2013;42:2294−320.

[16] Berger LI. In: Lide DR, editor. CRC handbook of chemistry and physics, vol. 88. Boca Raton, FL: CRC Press/Taylor and Francis; 2008.

[17] Emary C. Theory of nanostructures. 2009, p. 26 [Chapter 1].

[18] Tributsch H. Photovoltaic hydrogen generation. Int J Hydrogen Energy 2008;33:5911−30.

[19] Wang Z, Liu Y, Huang B, Dai Y, Lou Z, Wang G, et al. Progress on extending the light absorption spectra of photocatalysts. Phys Chem Chem Phys 2014;16:2758−74.

[20] Polman A, Atwater HA. Photonic design principles for ultrahigh-efficiency photovoltaics. Nat Mater 2012;11:174−7.

[21] Cowan AJ, Tang JW, Leng WH, Durrant JR, Klug DR. Water splitting by nanocrystalline TiO$_2$ in a complete photoelectrochemical cell exhibits efficiencies limited by charge recombination. J Phys Chem C 2010;114:4208−14.

[22] Tang JW, Durrant JR, Klug DR. Mechanism of photocatalytic water splitting in TiO$_2$. Reaction of water with photoholes, importance of charge carrier dynamics, and evidence for four-hole chemistry. J Am Chem Soc 2008;130:13885−91.

[23] Semonin OE, Luther JM, Choi S, Chen HY, Gao JB, Nozik AJ, et al. Nanocrystal solar cells squeeze extra juice out of sunlight. Science 2011;334:1530−3.

[24] Hashimoto K, Irie H, Fujishima A. TiO$_2$ photocatalysis: a historical overview and future propsects. Jpn J Appl Phys 2005;44:8269−85.

[25] Nakat K, Fujishima A. TiO$_2$ photocatalysis: design and applications. J Photochem Photobio C: Photochem Rev 2012;13:169−89.

[26] Nakata K, Ochiai T, Murakami T, Fujishima A. Photoenergy conversion with TiO$_2$ photocatalysis: new materials and recent applications. Electrochem Acta 2012;84:103−11.

[27] Gamage J, Zhang Z. Applications of photocatalytic disinfection. Int J Photoenergy 2010;2010:1−12.

[28] Blake DM, Maness PC, Huang Z, Wolfrum EJ, Huang J, Jacoby WA. Application of the photocatalytic chemistry of titanium dioxide to disinfection and the killing of cancer cells. Separ Purif Methods 1999;28:1−50.

[29] Ljubas D. Solar photocatalysis-a possible step in drinking water treatment. Energy 2005;30:1699−710.

[30] Basca R, Kiwi J, Ohno T, Albers P, Nadtochenko V. Preparation, testing and characterization of doped TiO$_2$ able to transform biomolecules under visible light irradiation by peroxidation/oxidation. J Phys Chem B 2005;109:5994−6003.

[31] Kern P, Schwaller P, Michler J. Electrolytic deposition of titania films as interference coatings on biomedical implants: microstructure, chemistry and nano-mechanical properties. Thin Solid Films 2006;494:279−86.

[32] Walther BA, Ewald PW. Pathogen survival in the external environment and the evolution of virulence. Bio Rev Camb Philos Soc 2004;79:849−69.

[33] Minero C, Maurino V, Pelizzetti E. Photocatalytic transformations of hydrocarbons at the sea water/air interface under solar radiation. Marine Chem 1997;58:361−72.

[34] Goswami DY, Vijayaraghavan S, Lu S, Tamm G. New and emerging developments in solar energy. Solar Energy 2004;76:33−43.

[35] Jacoby WA, Maness PC, Wolfrum EJ, Blake DM, Fennell JA. Mineralization of bacterial cell mass on a photocatalytic surface in air. Environ Sci Technol 1998;32:2650−3.

[36] Chen F, Yang X, Wu Q. Antifungal capability of TiO$_2$ coated film on moist wood. Build Environ 2009;44:1088−93.

CHAPTER 3

Heterogeneous Photocatalysts Based on Organic/Inorganic Semiconductor

3.1 INTRODUCTION

As we mentioned in Chapter 1, the heterogeneous photocatalysts based on different metal oxide and carbonaceous materials have shown better photocatalytic activity compared with that of single component. Various binary combinations of photocatalysts exhibited superior performance for decomposition of contaminants over a wide range. Prepared binary photocatalysts have been used successfully for detoxification of air as well as water for environmental remediation. Nevertheless, these efforts are not able to achieve desirable efficiency to commercialize application. The binary combinations suffer through low optical absorbance and large amount of recombination resulted in a poor performance. To further improve degradation efficiency and overcome the limitations of binary photocatalysts, attempts have been made to combine three different semiconductor nanostructures (ternary photocatalysts) with suitable band position and band gap energy. Ternary combinations show great potential to cover maximum portion of solar spectrum leading to better performance compared with that of binary photocatalysts. Therefore, in this chapter we will discuss binary and ternary photocatalysts prepared in our laboratory to degrade various contaminants.

3.2 BINARY PHOTOCATALYSTS

3.2.1 Sintering Assisted RGO/ZnO Composites for Water Purification Under UV Irradiation

As mentioned in Chapter 1, there are many sources of water pollution, including industrial wastes, untreated sewage, oxygen-demanding wastes, and inorganic and organic pollutants such as acids, salts, hydrocarbons, detergents, and toxic metals. Industrial waste and city sewage discharged into rivers are of particular concern. Industrial wastewater usually contains specific and readily identifiable chemical compounds. A large percentage of pollution is concentrated within a few subsectors, mainly in the form of toxic wastes and organic pollutants. Various methods have been frequently employed for water purification, including adsorption, biodegradation, electrocoagulation, nanofiltration, chlorination, ozonation, and advanced oxidation.

Among these methods, the advanced oxidation process using a heterogeneous photocatalyst is of special interest because it can degrade a wide range of organic pollutants. In particular, ZnO is one of the most promising catalysts because of its chemical stability, photosensitivity, fast electron transport capability, and better light absorption compared to that of TiO_2 [1]. ZnO has various important electrical and optical properties, such as high electron mobility at room temperature ($155 \, cm^2 \, V^{-1} \, s^{-1}$), high exciton binding energy (60 meV), and wide band gap energy (3.3 eV) [2]. However, its catalytic activity is poor because of efficient photoelectron recombination. Hence, attempts have been made to reduce photoelectron losses by making composites from metal nanoparticles (Ag, Pd, Au), metal oxides (TiO_2, SnO_2, Fe_2O_3), chalcogenides (ZnSe, CdS, ZnTe), and graphene [3]. Among these, graphene is the best choice to reduce photoelectron recombination and improve photodegradation efficiency. This is because its energy levels are suitable for photoelectron transfer and it has a high extraction ability for electrons from a semiconductor [4]. Many studies have investigated reduced graphene oxide/zinc oxide (RGO/ZnO) nanocomposites to reduce photoelectron recombination and improve photocatalytic activity [5]. However, reported nanocomposites have suffered from low photocatalytic efficiencies. Furthermore, the methods used for their fabrication required high temperatures, expensive substrates, vacuum systems, and rigorous experimental conditions and only provided the product in low yield. Preparing nanocomposites using a single reaction step at low cost and temperature, while maintaining good device performance, is highly desirable. The precursor sintering method is an excellent route to prepare oxide nanostructures in one step with high yield [6]. The synthesis does not require complex equipment or a vacuum system, thereby reducing the fabrication cost. Therefore, we reported sintering method to prepare graphene/ZnO nanocomposites for the application of water purification toward degradation of methylene blue (MB) under ultraviolet (UV) irradiation.

3.2.2 Preparation of RGO
The Hummers method was used to prepare graphene oxide (GO) powders [7]. In brief, 2 g of powdered graphite flakes was added with

stirring to 100 ml of sulfuric acid (H_2SO_4) that had been cooled below 10°C. Then, 8 g of potassium permanganate ($KMnO_4$) was gradually added, followed by 2 h of stirring at the same temperature. The temperature was raised to room temperature and the mixture stirred for an additional hour. The mixture was returned to a low-temperature bath and diluted with 100 ml of distilled water. Hydrogen peroxide (H_2O_2, 30%; 20 ml) was added to the mixture to dissolve any residual permanganate. A large amount of bubbles was released and the color of the mixture changed to brilliant yellow (Figure 3.1a). The colored suspension was filtered and washed several times with 1 M hydrochloric acid and distilled water. The retained GO powders were dried in an oven at 60°C for 12 h and stored in a vacuum oven.

The GO powders were reduced by dispersing 100 mg in 100 ml of distilled water with ultrasonication, adding 20 μl of hydrazine monohydrate drop by drop and refluxing the solution at 90°C for 2 h. The refluxed solution was filtered, and the retained RGO powders were transferred to a dish. The dish with the RGO was kept in an oven at 60°C for 12 h and then stored in a vacuum oven. The RGO powders were re-dispersed in distilled water under ultrasonication for 1 h; the black color (Figure 3.1b) of the solution indicated that the GO had been reduced completely.

Figure 3.1 Photographs of (a) graphene oxide in distilled water (0.5 mg/ml) and (b) reduced graphene oxide using hydrazine hydrate (0.5 mg/ml).

3.2.3 Preparation of RGO/ZnO Nanocomposites

RGO/ZnO nanocomposites were prepared from sintered precursors in a box furnace. Initially, the zinc source (5 g; i.e., zinc acetate (Zn $(CH_3COO)_2 \cdot 2H_2O$) or zinc nitrate ($Zn(NO_3)_2 \cdot 6H_2O$) powder) was sintered in a furnace in air at 450°C for 1 h. The nanocomposites were then fabricated by sintering the RGO powders (20 mg) and the zinc precursor (5 g) together in the furnace at 450°C for 1 h; the co-sintered powders were used directly for further analyses. The samples prepared with sintered zinc nitrate, zinc nitrate + RGO, zinc acetate, and zinc acetate + RGO are denoted as Z_N, Z_{NG}, Z_A, and Z_{AG}, respectively.

3.2.4 Photocatalytic Degradation of MB

The photocatalytic activity of the synthesized nanocomposites was measured by degrading MB (1.0×10^{-5} M, 200 ml) under UV light (300 W; Figure 3.2). An adsorption−desorption equilibrium of the photocatalyst and dye molecules was obtained by dispersing the nano-composite powder (0.005 g L^{-1}) in the dye solution by stirring at room temperature for 1 h in the dark. The dispersion was irradiated with the UV lamp (incident power = 50 mW cm^{-2}) while being stirred continuously. A 5-ml aliquot of the dispersion was withdrawn after each

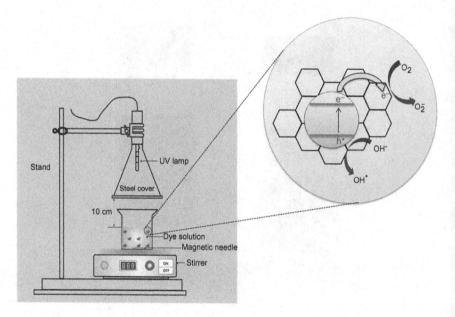

Figure 3.2 Schematic illustration of the experimental setup for photocatalytic process and enlarged image shows photo-electrons capture by graphene from ZnO conduction band to degrade dye molecules.

10 min of irradiation and centrifuged. The optical absorbance of the supernatant was measured using a spectrophotometer (V-600; Jasco, Tokyo, Japan). The dye concentration was monitored using the absorption band at 665 nm. The photocatalytic activities of commercial ZnO powders (Junsei Chemical, Tokyo, Japan) and the MB solution without catalyst were also measured for comparison under similar conditions.

3.2.5 Analysis of RGO/ZnO Composites

The crystal orientation and phase composition of sintered Z_N, Z_{NG}, Z_A, and Z_{AG} samples were analyzed using XRD (Figure 3.3a). Comparison of the observed patterns with the standard card (JCPDS No. 01-089-0511) indicated the hexagonal wurtzite crystal structure of ZnO aligned along the (101) direction. The patterns had peaks corresponding to the (100), (002), (102), (110), (103), (200), and (201) planes, indicating the polycrystalline nature of the ZnO structure. The pattern for RGO is provided for comparison. The existence of two peaks at 2θ values of 20.6° and 42.6° in the RGO XRD pattern confirmed the complete reduction of graphene oxide by hydrazine hydrate.

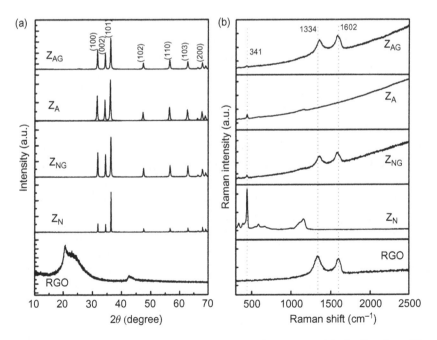

Figure 3.3 (a) XRD patterns and (b) Raman spectra in the range of 300–2500 cm^{-1} for Z_N, Z_{NG}, Z_A, and Z_{AG} samples. Spectrum of pure RGO is given for reference.

However, graphene peaks were not seen in the Z_{NG} and Z_{AG} XRD patterns due to the small amount of graphene sheets that had been co-sintered with the zinc precursors (a quantity that was too small to be detected by XRD). Careful analysis of the XRD patterns revealed very small changes in crystallinity after the addition of graphene, which was evidence for the formation of the RGO/ZnO nanocomposites.

The quality of the RGO/ZnO nanocomposites was studied using Raman spectroscopy (Figure 3.3b). The Raman spectrum for RGO was used to clarify the Raman shifts of the bands observed for the nanocomposites. The two strong peaks observed for RGO at about $1334\ cm^{-1}$ (D band) and $1602\ cm^{-1}$ (G band) are attributable to disordered sp^2-bonded carbon and to breathing vibrations of six-membered sp^2 carbon rings (E_{2g} and A_{1g} modes) in reduced graphene [8]. In case of the Z_N and Z_A samples, a single peak at about $442\ cm^{-1}$ is the optical phonon E2 (transverse optical) in the Brillouin zone of the ZnO nanostructure. However, the Z_{NG} and Z_{AG} XRD patterns contained peaks for RGO and ZnO at values shifted toward a lower wave number ($439\ cm^{-1}$), indicating a strong interaction between the RGO nanosheets and ZnO nanostructures.

The surface morphology and growth of the ZnO nanostructures with and without RGO were studied using FE-SEM. A well-faceted, micrometer-scale pyramid-like structure was formed using zinc nitrate sintering (Figure 3.4a), with faceted growth uniformly distributed throughout the sample. However, in the presence of RGO, the surface morphology changed from a faceted structure to one of smooth crystals in which graphene sheets were mixed with ZnO crystals. The low-magnification image shows an uneven distribution of ZnO crystals and RGO and poor sample homogeneity. The formation of the pyramid-like structures with nitrate sintering is attributable to nitrate ions from the zinc nitrate. These ions adsorb on the $(10\bar{1}1)$ plane, which retards growth and results in pyramid-like structures. In the case of zinc acetate sintering, nanorods formed with an average diameter of 60 nm. The low-magnification image shows a uniform distribution of the nanorods. For RGO and zinc acetate sintering, the nanoparticles, with an average diameter of 70 nm, were well distributed on the graphene sheet surfaces. Uniform coverage of nanoparticles over an entire graphene sheet is evident in the low-magnification image. The RGO/ZnO nanocomposite was also studied using high-resolution transmission

Figure 3.4 (a) FESEM images of Z_N, Z_{NG}, Z_A, and Z_{AG} samples and (b) TEM image for Z_{AG} sample showing the uniform distribution of ZnO nanoparticles with an average diameter of 70 nm over entire graphene sheet.

electron microscopy (HR-TEM) (Figure 3.4b). The TEM image reveals a uniform distribution of the ZnO particles over the graphene sheet. The nanoparticles are well distributed on the graphene surfaces because of strong electrostatic and electron transfer interactions between ZnO and graphene.

The photocatalytic activity of the RGO/ZnO nanocomposites was measured for degradation of MB under UV irradiation. Photocatalytic

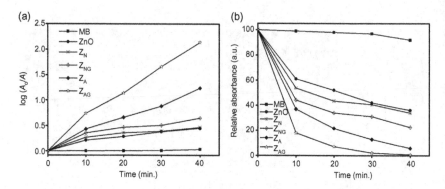

Figure 3.5 (a) Relative absorbance and (b) logarithmic plots for commercial ZnO powders, Z_N, Z_{NG}, Z_A, and Z_{AG} samples with respect to UV irradiation time. The absorbance of MB without photocatalysis is also shown for comparison.

efficiency was assessed by monitoring the change in MB absorbance over time. The change in relative absorbance of the MB solution with time in the presence of different catalysts under UV light irradiation is shown in Figure 3.5a, where the relative absorbance is $A_0/A \times 100$, where A is the absorbance of the MB solution at the irradiation time t and A_0 is the absorbance of the solution before irradiation ($t = 0$). A slight decrease in the MB concentration was found under UV irradiation in the absence of any catalyst. However, photodegradation occurred after the addition of commercial ZnO powder, with a catalytic efficiency of 64%. Graphene incorporation with Z_N to form Z_{NG} improved the efficiency from 66% to 78.6%. The photodegradation efficiency improved from 94.3% to 99.3% when graphene was sintered with Z_A to form Z_{AG}. Hence, graphene addition improved the catalytic efficiency in all cases. Figure 3.5b shows logarithmic graphs $\ln(A_0/A) = -kt$, where A_0/A is the normalized absorbance of the MB solution, k is the apparent kinetic rate constant, and t is the reaction time. This linear $\ln(A_0/A)$ versus t plot demonstrates that the photodegradation of MB follows pseudo first-order reaction kinetics. The calculated value of k for the Z_{AG} sample ($k = 52.10 \times 10^{-3}\,\mathrm{min}^{-1}$) was four times greater than that for commercial ZnO powder ($k = 10{,}050 \times 10^{-3}\,\mathrm{min}^{-1}$). The k values of Z_A ($k = 29.20 \times 10^{-3}\,\mathrm{min}^{-1}$), Z_N ($k = 10.60 \times 10^{-3}\,\mathrm{min}^{-1}$), and Z_{NG} ($k = 14.40 \times 10^{-3}\,\mathrm{min}^{-1}$) samples were slightly higher than that of commercial ZnO powder but lower than that of the Z_{AG} sample (Table 3.1). Hence, the nanocomposites prepared with RGO and zinc acetate sintering, that is, Z_{AG}, had better photocatalytic activity than the Z_N, Z_{NG}, and Z_A samples.

Table 3.1 Summary of the Measured Specific Surface Area (m^2 g^{-1}) and Reaction Rate Constants (min^{-1}) for Commercial ZnO Powders, Z$_N$, Z$_{NG}$, Z$_A$ and Z$_{AG}$ Samples

Sample Details	BET Surface Area (m^2 g^{-1})	Kinetic Rate Constant (min^{-1}) $\times$ 10^{-3}
Zinc nitrate (Z$_N$)	2.09	10.60
Zinc nitrate-Graphene (Z$_{NG}$)	4.79	14.40
Zinc acetate (Z$_A$)	11.50	29.20
Zinc acetate-Graphene (Z$_{AG}$)	12.83	52.10

Basically, the photocatalytic activity depends on the effective separation of electron−hole pairs, specific surface area, and crystallinity. In the present work, synthesized nanocomposites exhibited excellent photocatalytic activity compared with that of pristine ZnO nanostructures. Their enhanced photocatalytic activities is attributable to the high effective surface area (e.g., 12.83 m^2 g^{-1} for Z$_{AG}$) and efficient separation of photoelectrons because graphene acts as an electron acceptor (see the enlarged schematic in Figure 3.2). Additionally, the strong $\pi - \pi$ stacking interactions between MB dye molecules and the graphene surface increases the reactivity.

3.2.6 Possible Photocatalysis Mechanism

The enhanced photocatalytic activities of the RGO/ZnO composites stem from suitable energy band positions of ZnO (-4.05 eV) and graphene (-4.42 eV). This improves photoelectron transfer, and the high electron mobility of graphene transfers the injected electrons quickly from the interface to the sheets, minimizing recombination losses. Hence, photocatalytic activity improved significantly with RGO addition to ZnO. The following series of reaction processes involved in the MB degradation are proposed:

$$ZnO + h\nu(\geq 3.3 \text{ eV}) \rightarrow ZnO(h_{VB}^+ + e_{CB}^-) \tag{3.1}$$

$$ZnO(h_{VB}^+ + e_{CB}^-) + RGO \rightarrow ZnO(h_{VB}^+) + RGO(e_{CB}^-) \tag{3.2}$$

$$h_{VB}^+ + H_2O \rightarrow H^+ + OH^\bullet \tag{3.3}$$

$$H^+ + OH^- \rightarrow \cdot OH \tag{3.4}$$

$$RGO(e_{CB}^-) + O_2 \rightarrow \cdot O_2^- \tag{3.5}$$

$$\cdot O_2^- + HO_2^\bullet + H^+ \rightarrow H_2O_2 + O_2 \tag{3.6}$$

$$H_2O_2 + \cdot O_2^- \rightarrow \cdot OH + OH^- + O_2 \qquad (3.7)$$

$$H_2O_2 + e_{CB}^- \rightarrow \cdot OH + OH^- \qquad (3.8)$$

Under UV light with an energy ≥ 3.3 eV, electron–hole pairs are generated, and then photogenerated electrons are transferred to RGO nanosheets as described in Eqs. (3.1) and (3.2). The holes in $ZnO(h_{VB}^+)$ generate OH radicals as shown in Eqs. (3.3) and (3.4). The electrons captured on the RGO react with oxygen, which forms transient superoxide radicals (Eq. (3.5)). Finally, the superoxide molecules react with electrons in the RGO (Eq. (3.6)) and form highly reactive OH radicals, which mineralize MB molecules as shown in Eqs. (3.7) and (3.8). Thus, graphene plays a crucial role in reducing recombination losses and thereby increases photocatalytic activity.

3.2.7 Visible Light Active RGO/CdS Heterojunctions for Cr(VI) Reduction

To further improve photocatalytic performance and range of degradation, we reported fabrication of RGO/CdS composites synthesized through chemical bath deposition (CBD) method. It is known that graphene is the most important form of carbon due to its various fascinating electrical, optical, and mechanical properties [9]. It can be used in several applications such as photocatalysis, solar cells, supercapacitor, batteries, field emission, antibacterial activity, fuel cells, chemical detectors, and many others [10]. Particularly, it is most suitable in photocatalysis because of its strong electron capture ability and excellent transportation capability, and high specific surface area for adsorption of organic pollutants [11]. Hence, it has been combined with various semiconductors such as TiO_2, ZnO, SnO_2 Fe_3O_4, $BiVO_4$, and CdS using hydrothermal method, microwave assisted method, solvothermal method, and anodization method for the photodegradation of organic pollutants under UV and visible light [12]. The aforementioned semiconductors and graphene nanocomposites have been used successfully to degrade several toxic and hazardous pollutants such as lead, mercury, cadmium, and hexavalent chromium (Cr(VI)), presented in wastewater. Among these pollutants, Cr(VI) is known to be highly toxic due to its carcinogenic property in hexavalent form [13]. Various research groups have been investigated regarding Cr(VI) reduction into Cr(III), which is a less toxic phase. Although these groups successfully reduced Cr(VI) using the above-mentioned techniques, the main limitation of these catalysts is its poor performance. Further, few of them used hydrogen peroxide and UV light for Cr(VI) reduction, which is not suitable for the sunlight. Hence, it is necessary to fabricate grapheme-based composites with visible band gap semiconductors for excellent reduction of Cr (VI). Hence, we explored deposition of RGO-CdS composites using CBD method for reduction of Cr(VI) under visible light.

3.2.8 RGO/CdS Fabrication

For the fabrication of RGO/CdS nanocomposites, 0.20 wt% of RGO powders were dispersed in distilled water using ultrasonication for 1 h. Here, 0.20 wt% of RGO powders is the optimized ratio between RGO and CdS. Then cadmium sulfate, thiourea, and ammonia solution were added as mentioned above under constant stirring. Subsequently, cleaned glass/FTO (fluorine-doped tin oxide-coated glass substrate was dipped into dispersed RGO and cadmium solution at 70°C for 5 h. Finally, the deposited substrate was removed and rinsed in distilled water, and dried in air at room temperature.

3.2.9 Analysis of RGO/CdS

Detailed surface morphology of RGO/CdS samples was evaluated using TEM. As shown in Figure 3.6a, it is found that spherical shaped CdS nanoparticles are distributed onto the surface of graphene sheet. All nanoparticles are found to be covered with CdS nanoparticles completely. All spherically shaped CdS nanoparticles with ~30 nm in diameter are distributed onto graphene sheets without any aggregation. The RGO/CdS composite was also studied using high-resolution transmission electron microscopy (HR-TEM) (Figure 3.6b). From HR-TEM analysis, the d-spacing was found to be 0.33 nm, oriented along (002) plane and this result is consistent with XRD analysis. The nanoparticles are distributed on the graphene surface because of strong electrostatic and electron transfer interactions between CdS and graphene sheet. The selected area electron diffraction (SAED) pattern indicates that the RGO/CdS sample (inset of Figure 3.6b) is a polycrystalline structure of CdS nanoparticles. The SAED pattern exhibits three rings oriented along (002), (100), and (101) planes of the hexagonal phase for CdS nanoparticles, respectively. The elemental composition of RGO/CdS sample was determined using spot energy dispersive X-ray spectroscopy (EDX). The EDS results (Figure 3.6c) give signals of C, Cd, and S elements, which confirm the successful formation of RGO/CdS nanocomposite.

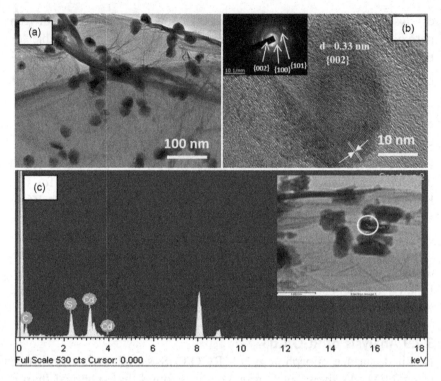

Figure 3.6 TEM image of (a) CdS nanoparticles coated over entire RGO sheet, (b) HR-TEM image with an interplanar spacing of 0.33 nm and orientation along the (002) direction, and (c) EDS spectra of RGO-CdS sample.

Raman spectroscopy is the best nondestructive tool to determine the quality of crystalline materials and formation of nanocomposites. The RGO, CdS, and RGO/CdS samples were studied using Raman spectroscopy (Figure 3.7). The Raman spectrum for RGO was used to clarify the Raman shifts of the bands observed for the nanocomposites. The existence of two strong peaks for RGO at about $1334\,\text{cm}^{-1}$ (D band) and $1602\,\text{cm}^{-1}$ (G band) is attributed to disordered sp^2-bonded carbon and to breathing vibrations of six-membered sp^2 carbon rings (E_{2g} and A_{1g} modes) in reduced graphene. In addition to these peaks, there is one more low intensity peak at $\sim 2670\,\text{cm}^{-1}$ (2D) due to double phonon resonance process. The variation in 2D peak can reveal a number of stacking layers of graphene [14]. The lower intensity of 2D peak indicates that the synthesized graphene sheets are stacked with few layers of graphene. For CdS sample, Raman spectra showing the two peaks at $304\,\text{cm}^{-1}$ and $602\,\text{cm}^{-1}$ correspond to longitudinal

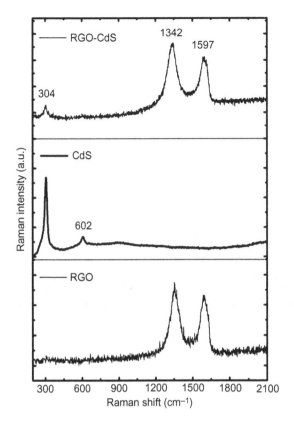

Figure 3.7 Raman spectra in the range of 200–2100 cm^{-1} for CdS and RGO-CdS samples. Spectrum of pure RGO is given for reference.

optical phonon modes [15]. Finally, in case of RGO/CdS sample, a single peak at about 304 cm^{-1} is the optical phonon LO mode of CdS, and two more peaks at 1334 cm^{-1} and 1602 cm^{-1} correspond to RGO. Hence, the existence of peaks related to RGO and CdS indicate the successful formation of RGO/CdS nanocomposite.

3.2.10 Cr(VI) Reduction

The photocatalytic reduction of Cr(VI) was evaluated by measuring the absorbance from UV-visible spectrophotometer. Photocatalytic performance was estimated by monitoring the change in absorbance over irradiation time for potassium dichromate solution at 372 nm. The change in absorbance of the potassium dichromate solution with time in the presence of different catalysts under visible light irradiation is shown in

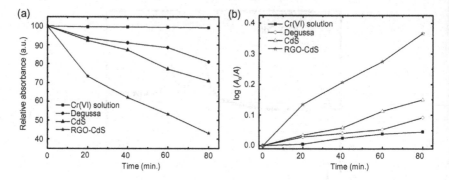

Figure 3.8 (a) Relative absorbance for Degussa powders, CdS, and RGO–CdS samples with respect to visible light irradiation time. The absorbance of Cr(VI) solution under visible light irradiation without catalysis is also shown for comparison and (b) the ln(A₀/A) versus t curves of Cr(VI) photocatalytic reduction without catalyst and with CdS and RGO–CdS catalysts.

Table 3.2 Summary of the Measured Specific Surface Area (m² g⁻¹) and Reaction Rate Constants (min⁻¹) for Degussa Powders, CdS, and RGO–CdS Samples

Sample Details	BET Surface Area (m² g⁻¹)	Kinetic Rate Constant (min⁻¹) × 10⁻³
Cr(VI) solution	–	0.06
Degussa powders	18–20	1.04
CdS	2.57	1.88
RGO-CdS	47.44	4.36

Figure 3.8a. It is seen that Cr(VI) cannot be reduced under visible light without photocatalyst and a slight reduction of Cr(VI) ions was found for Degussa powders with its reduction rate of 11.39%. However, photocatalytic reduction rate occurred faster after the addition of CdS powders, with its catalytic efficiency of 29.11%. After incorporating RGO with CdS to make nanocomposites, the photocatalytic reduction rate improved almost two times, that is, 56.96%. Hence, adding graphene significantly improved the photocatalytic reduction efficiency. As shown in Figure 3.8b, the linear $\ln(A_o/A)$ versus t plot revealed that the reduction of Cr(VI) ions follows pseudo first-order reaction kinetics [16]. The calculated value of k for the RGO/CdS sample ($k = 4.36 \times 10^{-3}$ min⁻¹) was three times higher than that for Degussa powders ($k = 1.04 \times 10^{-3}$ min⁻¹). The k value of CdS sample ($k = 1.88 \times 10^{-3}$ min⁻¹) was slightly higher than that of Degussa powders but lower than that of the RGO/CdS sample (Table 3.2). Hence, the nanocomposites prepared with RGO and CdS exhibited better photocatalytic reduction of Cr(VI) ions than those using the Degussa powders and CdS samples. The observed

photocatalytic activity is much higher than those reported values for other RGO/CdS nanocomposites in the literatures [17].

In general, the photocatalytic reduction of organic pollutants mainly depends on the effective transfer of photoelectrons and specific surface area of nanostructure. The improved reduction of Cr(VI) ions is attributed to effective photoelectrons transfer and high specific surface area of RGO/CdS sample.

3.2.11 Photoreduction Performance
The photocatalytic performance of catalysts basically depends on (i) optical absorption and (ii) effective transportation of photoelectrons, which hinder the electron−hole recombination. The optical absorbance of CdS and RGO/CdS samples has been measured using UV-Visible spectrophotometer in the range of 350−900 nm (Figure 3.9a). Both samples exhibited absorption edge at 530 nm, which is assigned to band gap absorption of CdS nanoparticles. Hence, addition of graphene does not change the band gap energy of CdS nanoparticles. However, the optical absorbance for RGO/CdS sample increased compared with that for CdS samples in the visible region. The increased absorption is due to the addition of graphene. Inset of Figure 3.9a shows optical images of CdS and RGO/CdS samples. It is seen that there is no noticeable change in the color of sample after addition of graphene. Therefore, improved absorbance was found to be one of the reasons for effective photocatalytic reduction of Cr(VI) under visible light.

The suitable energy levels of RGO and CdS are playing crucial role for fast photoelectrons transfer. The conduction band position of CdS (-3.6 eV) and graphene work function (-4.42 eV) are beneficial for effective separation of photogenerated electron−hole pairs [18]. This improves photoelectron transfer, and the high electron mobility of graphene transfers the injected electrons quickly from the interface to the sheets, minimizing recombination losses (schematic shown in Figure 3.9b). Hence, photocatalytic activity improved significantly with RGO addition to CdS. The following series show reaction processes involved in the reduction of Cr(VI) to Cr(III) by photoelectrons [19,20]:

$$CdS + h\nu(\geq 2.42 \text{ eV}) \rightarrow CdS(h_{VB}^{+} + e_{CB}^{-}) \tag{3.9}$$

$$CdS(h_{VB}^{+} + e_{CB}^{-}) + RGO \rightarrow CdS(h_{VB}^{+}) + RGO(e_{CB}^{-}) \tag{3.10}$$

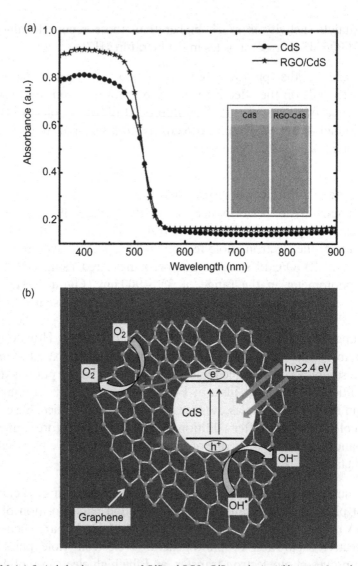

Figure 3.9 (a) Optical absorbance spectra of CdS and RGO−CdS samples in visible region. Inset shows photographs of CdS and RGO−CdS samples, and (b) schematic diagram of RGO−CdS sample shows the electron transfer mechanism under visible light and photocatalytic reduction of Cr(VI).

$$Cr_2O_7^{2-} + 14H^+ + 6e^- \rightarrow 2Cr^{3+} + 7H_2O \qquad (3.11)$$

$$2H_2O + 4h^+ \rightarrow O_2 + 4h^+ \qquad (3.12)$$

Under visible light with an energy ≥2.42 eV, electron−hole pairs are generated, and then photogenerated electrons are transferred to RGO nanosheets as described in Eqs. (3.9) and (3.10). After electron−hole pairs

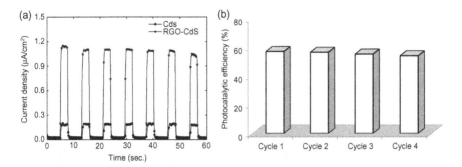

Figure 3.10 (a) Photocurrent transient responses of CdS and RGO/CdS samples recorded under visible light in 0.5 M Na₂SO₄ electrolyte and (b) photocatalytic stability of RGO/CdS sample over four cycles.

are separated, the photoelectrons can reduce Cr(VI) to Cr(III) as shown in Eqs. (3.11) and (3.14). Thus, graphene plays a crucial role in the separation of photoelectrons, leading to low recombination losses and thereby increasing photocatalytic reduction of Cr(VI) to Cr(III) under visible light.

To confirm the role of RGO in photocatalytic improvement, photocurrent transient responses were measured for CdS and RGO/CdS samples. Figure 3.10a shows photocurrent–time (I–t) response of CdS and RGO/CdS samples under intermittent illumination of visible light. It was found that both electrodes showed good reproducibility and stability after a number of cycles. The photocurrent became zero once lamp was off and came back close to original value once lamp was on. Interestingly, the photocurrent density of RGO/CdS sample (1.13 mA cm^{-2}) is significantly higher than that of CdS sample (0.20 mA cm^{-2}). The improvement in photocurrent is attributed to electron capture and transfer ability of RGO, resulting in high photocatalytic performance of RGO/CdS compared with that of CdS sample only. Photocatalytic stability is one of the important factors in photocatalysis for the practical use of fabricated photocatalysts. Hence, in this study, authors have investigated the stability of RGO/CdS sample over four cycles. As shown in Figure 3.10b, the photocatalytic efficiency of RGO/CdS sample remains almost the same up to four cycles. Therefore, synthesized RGO/CdS composite is stable and suitable for practical application.

3.3 TERNARY PHOTOCATALYSTS

3.3.1 Nanocomposites Based on RGO, CdS, and ZnO

Heterogeneous photocatalysts play crucial roles in advanced water purification devices because of their broad applicabilities and better performances compared to homogeneous catalysts. Among the photo-catalysts, ZnO has been shown to be promising because of its various interesting properties. Despite their good performances and stabilities, ZnO-based photocatalysts have limited applications because of a wide band gap energy (3.37 eV) and high recombination rate. Hence, various research groups have tried to improve its photocatalytic activity using different approaches such as element doping, sensitization with a visible band gap semiconductor, and use of gold (Au) nanoparticles [21]. However, these combinations suffered from low catalytic performance, low visible light absorptivity and stability, and high cost.

Recently, low band gap semiconductor sensitization has been reported as the best option spanning the UV as well as the VIS portions of the solar spectrum [22]. A low band gap semiconductor can potentially use multiple electron–hole pair generation per incident photon to achieve higher photocatalytic activity [23]. Various researchers have proposed visible-band gap semiconductors for effective sensitization, such as CdSe, InP, CdTe, PbS, and CdS [24]. Among these sensitizers, CdS is highly promising because of its reasonable band gap (2.42 eV), which offers new opportunities for light harvesting. Although reported combination of photocatalysts had satisfactory photodegradation performance, it was not sufficient for commercial photocatalytic devices. Improving the photocatalytic efficiency is essential by avoiding recombination losses and promoting fast electron transportation.

With the discovery of RGO, a wide range of potential applications was expected because of its remarkable properties, including high electron mobility at room temperature, large theoretical specific surface area $(2630 \, m^2 \, g^{-1})$, excellent thermal conductivity $(3000–5000 \, Wm^{-1} \, K^{-1})$, good optical transparency (97.7%), and high Young's modulus (~1 TPa) [25]. RGO has been combined with various semiconductors, including ZnO, CdS, TiO_2, Fe_3O_4, SnO_2, Cu_2O, and WO_3, and its use has been explored in supercapacitors, solar cells, gas sensors, batteries, and photocatalysts [26,27]. It plays a crucial role in photocatalysis for the effective separation of photogenerated electron–hole pairs because of its electron-capturing ability [28]. We investigated the incorporation of RGO into CdS nanoparticles (CNPs) and ZnO nanorods for the effective photodegradation of MB dye in visible light. A unique low-temperature, water-based method for the single-step sensitization of RGO and CdS on ZnO nanorods is described in this section.

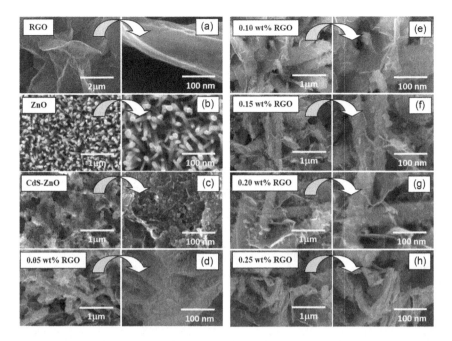

Figure 3.11 FE-SEM images of (a) RGO sheets, (b) Vertically aligned ZnO nanorods, (c) CdS-sensitized ZnO nanorods, (d) RGO- and CdS-sensitized ZnO nanorods with 0.05 wt% of RGO, (e) RGO−CdS−ZnO composite with 0.10 wt% of RGO, (f) RGO−CdS−ZnO composite with 0.15 wt% of RGO, (g) RGO−CdS−ZnO composite with 0.20 wt% of RGO, and (h) RGO−CdS−ZnO composite with 0.25 wt% of RGO at ×50 K and ×150 K.

3.3.2 Analysis of RGO, CdS, and ZnO Composites

The surface morphologies of RGO, ZnO nanorods, CdS-sensitized ZnO nanorods, and composites of RGO−CdS−ZnO were examined using FESEM as a function of RGO content. RGO sheets several micrometers in size were formed after the reduction of graphene oxide (Figure 3.11a). The magnified image shows a thickness of 70 nm, which indicates stacking of graphene sheets. Figure 3.11b is a FE-SEM image of ZnO nanorods chemically grown on a glass substrate. The smooth and aligned nanorods grew over the entire substrate and had an average diameter of about 70 nm (magnified image of Figure 3.11b). After CdS sensitization, the surface morphology changed significantly (Figure 3.11c). The nanorods were completely coated with 30-nm-diameter CNPs; the surface was rougher than that of bare ZnO nanorods. All of the nanorods were covered with CNPs, indicating successful sensitization of the particles (magnified image in Figure 3.11c). FE-SEM images of ZnO nanorods sensitized with different amounts of RGO and CNPs are shown in Figure 3.11c−g.

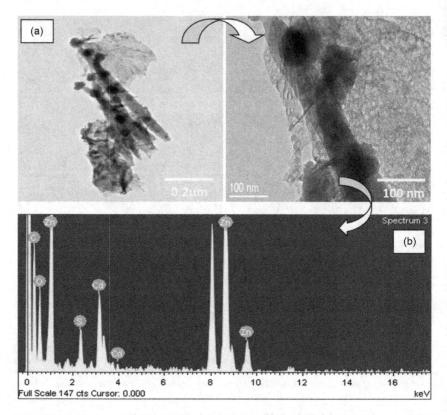

Figure 3.12 (a) TEM image of RGO- and CdS-sensitized ZnO nanorods with 0.20 wt% of RGO and its magnified image showing CdS particles attached to ZnO nanorods and (b) EDS spectra of the RGO–CdS–ZnO composite.

The RGO and CNPs were deposited together onto the surface of the ZnO nanorods so that the nanorods were completely covered. Wrapping of the nanorods took place as the amount of RGO increased from 0.05 to 0.25 wt%. The number of RGO sheets increased with the addition of 0.25 wt% of RGO to the sample (Figure 3.11h). More RGO sheets may improve electron transport while simultaneously lowering the photoelectron recombination rate.

The microstructures of the RGO–CdS–ZnO composites were further studied using TEM (Figure 3.12a). It shows that CNPs were distributed over entire RGO sheets as well as they tethered to ZnO nanorods. Further, the magnified TEM image confirms that CNPs (~ 50 nm) were attached to ZnO nanorods as well as RGO sheets. The elemental composition of an RGO–CdS–ZnO composite was

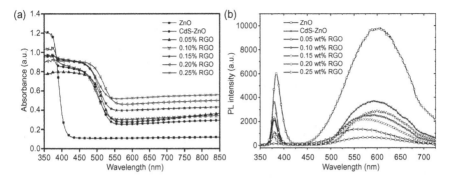

Figure 3.13 (a) Optical absorbance and (b) PL spectra at room temperature of ZnO, CdS–ZnO, and RGO–CdS–ZnO composites.

determined using spot energy dispersive X-ray spectroscopy (EDS). The EDS results (Figure 3.12b) showed Cd, S, Zn, and O, which was consistent with CdS and ZnO being present in the RGO–CdS–ZnO composite.

High optical absorbance is essential for effective photocatalysis. In particular, absorbance in the visible region is important to cover a large portion of the solar spectrum. The absorbance spectra of ZnO films with and without RGO–CdS composites were recorded using a UV–VIS spectrophotometer in the range of 350–850 nm. Figure 3.13a shows that a significant increase in absorbance in the UV region for ZnO nanorods can be attributed to intrinsic band gap energy absorption (3.3 eV). After CdS sensitization, the absorption edge shifts toward the visible region because of the band gap energy of the CNPs (2.4 eV), and its absorbance in the visible region increased after CdS sensitization compared with a ZnO film. RGO–CdS–ZnO films with increasing RGO content showed further increments in UV as well as visible absorbances, while the absorption edge remained at a similar position. Among them, the composite with 0.25 wt% of graphene had the highest absorbance in the visible region. This indicated that the surfaces of the ZnO nanorods were successfully sensitized by the RGO and CNPs. The enhanced absorbance of composites with increasing RGO content indicated the contribution of RGO to the absorbance. These results agree with data reported elsewhere [29]. Finally, the increased absorbance in UV and visible regions will improve light harvesting capacity and should enhance the photocatalytic performance of composite samples compared with those made with only CNPs and ZnO nanorods.

Faster electron transfer from the conduction band of the metal oxide to RGO can minimize the recombination rate. The PL spectra (Figure 3.13b) have both UV and VIS emission peaks. The UV peak (~381 nm) was attributed to near-band-edge transition processes arising from energy loss due to strong electron–phonon interactions at room temperature. The visible peak (~600 nm) was present in all samples but changed in intensity. The high intensity of the visible peak for the ZnO nanorods (~10,000) indicated a high recombination rate. However, the rate decreased with sensitization by CNPs and RGO sheets, and was found to be the lowest for the composite with 0.25 wt% RGO (~1000). This tenfold lower intensity indicated that PL quenching by the RGO had occurred, which would reduce the recombination rate [30]. Therefore, RGO incorporation was expected to increase the photocatalytic degradation rate.

3.3.3 Photocatalysis of RGO, CdS, and ZnO Composites

The photocatalytic activities toward MB under visible illumination were evaluated under the same experimental conditions for the synthesized RGO–CdS–ZnO composites. The photocatalytic performance of MB without the catalyst was also measured as a control. The relative variation in absorbance of the MB solution with respect to time in the presence of different photocatalysts under visible light illumination is shown in Figure 3.14a. Irradiation by visible light in the absence of catalyst caused a slight decrease in the MB concentration. However, photodegradation of MB occurred significantly after its addition to CdS–ZnO powders and to different RGO–CdS–ZnO composites. The photodegradation rate of the composite with 0.20 wt% RGO was higher than those observed for the other composite samples. Therefore,

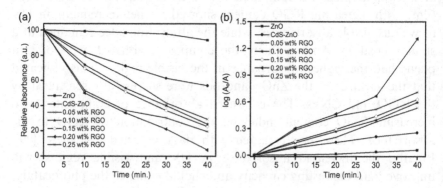

Figure 3.14 (a) Relative optical absorbance and (b) logarithmic ln(A₀/A) versus t curves for MB without photocatalyst and CdS–ZnO and RGO–CdS–ZnO composites as a function of irradiation time.

Table 3.3 Summary of the Measured Specific Surface Area (m^2g^{-1}) and Reaction Rate Constants (min^{-1}) for Degussa Powders, CdS, and RGO–CdS Samples

Sample Details	Specific Surface Area (m^2g^{-1})	Kinetic Rate Constant (K) min^{-1}
MB	–	0.001
CdS-ZnO	14.38	0.006
0.05 wt% RGO	14.69	0.012
0.10 wt% RGO	15.15	0.014
0.15 wt% RGO	15.72	0.016
0.20 wt% RGO	16.83	0.028
0.25 wt% RGO	23.08	0.018

these differences in degradation rate indicated that the photodegradation depended on the percentage of RGO in the composite sample.

The data in Figure 3.14b were also interpreted using the equation ln $(A_0/A) = -kt$; the linear plots revealed that the photodegradation of MB followed pseudo first-order reaction kinetics. The calculated value of k for the 0.20 wt% RGO–CdS–ZnO ($k = 0.028\,min^{-1}$) sample was almost four times higher than that for CdS–ZnO and other composite samples. The k value for the 0.25 wt% RGO–CdS–ZnO sample ($k = 0.018\,min^{-1}$) was slightly higher than that for the 0.05 ($k = 0.012\,min^{-1}$), 0.10 ($k = 0.014\,min^{-1}$), and 0.15 wt% ($k=0.016\,min^{-1}$) samples, but was lower than that for the 0.20 wt% sample (Table 3.3). Hence, the RGO–CdS–ZnO composite containing 0.20 wt% RGO had better photocatalytic performance than the samples synthesized with different percentages of RGO and without RGO. Photodegradation performance mainly depends on photoelectron separation before recombination and rapid transport. The improved catalytic performance found for the 0.20 wt% RGO–CdS–ZnO sample is the result of their effective separation and transport because of a PL quenching effect, increased optical absorbance, and specific surface area. However, the photocatalytic performance deteriorated beyond the optimum amount of RGO. The obtained results can be attributed to the wrapping of CdS and ZnO nanorods by RGO sheets, so that large amount of sheets may act as recombination centers.

3.3.4 Possible Electron Transport Mechanism
The enhanced photocatalytic activities of the RGO–CdS–ZnO composites stem from suitable energy band positions of RGO (−4.42 eV), ZnO (−4.21 eV), and CdS (−3.98 eV) (Figure 3.15). This improved

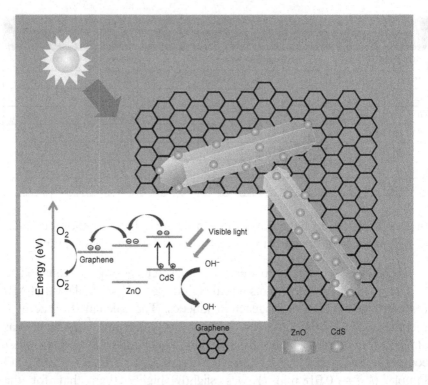

Figure 3.15 Schematic diagram of a RGO−CdS−ZnO composite showing the electron transfer mechanism under visible light and photocatalytic degradation of MB dye.

photoelectron transfer and the high electron mobility of graphene transferred the injected electrons quickly from the interface to the sheets, minimizing recombination losses. Hence, photocatalytic activity improved significantly with the addition of RGO to the CdS−ZnO composite. A proposed series of reaction processes involved in the MB degradation is as follows:

$$CdS + ZnO + graphene \xrightarrow{h\nu} CdS(h_{VB}^+) + ZnO(e_{CB}^-) + graphene(e^-) \quad (3.13)$$

$$CdS(h_{VB}^+) + H_2O \rightarrow CdS + OH^{\bullet} \quad (3.14)$$

$$graphene(e^-) + O_2 \rightarrow graphene + O_2^- \quad (3.15)$$

Electron−hole pairs are generated under visible light illumination with an energy ≥ 2.42 eV. Photogenerated electrons are then transferred toward the ZnO conduction band and to RGO sheets (Eq. 3.13). The holes in $CdS(h_{VB}^+)$ generate $OH^{\bullet}$ radicals (Eq. 3.14). The electrons captured on the RGO react with oxygen to form transient superoxide radicals and superoxide molecules (Eqs. 3.15 and 3.6). Highly reactive OH$^{\bullet}$ radicals are formed, which mineralize MB molecules as shown in Eqs. 3.7 and 3.8. Thus, graphene plays a crucial role in reducing recombination losses and thereby increasing photocatalytic activity.

3.4 NANOCOMPOSITES BASED ON RGO, CNTs, AND Fe$_2$O$_3$

There is growing demand for green technologies to reduce the environmental impact of modern industry and to develop cleaner energy sources. Semiconductor photocatalysis is one of the most attractive technologies for applications, including degradation of a wide range of pollutants, water splitting for H$_2$ production, and transformation of volatile organic compounds into biodegradable molecules [31]. Various semiconductors, including TiO$_2$, ZnO, WO$_3$, SnO$_2$, Fe$_2$O$_3$, CdS, and CdSe, have been employed for applications in water purification and H$_2$ production [32]. Hematite (Fe$_2$O$_3$) is one of the most widely studied semiconductor photocatalysts because of its low environmental impact, corrosion resistance, and low cost [33]. It is a strong magnetic material, which facilitates separation of the photocatalyst from a polluted liquid so that it can be reused. Moreover, it has an optical band gap of 2.1 eV, which is in the visible part of the electromagnetic spectrum, meaning that photocatalytic degradation is possible under irradiation with sunlight [34]. However, the photocatalytic performance of Fe$_2$O$_3$ is not yet sufficient for commercial or water purification applications. The performance of photocatalysts depends on several factors, including the crystallinity of the material, the size and shape of the nanostructure, the specific surface area, the pollutant adsorption capacity, the optical absorption, and the recombination rate of electron−hole pairs. A number of approaches have been used to grow hierarchical nanostructures, as well as composites with other semiconductors and conducting materials (including carbon nanotubes (CNTs), graphene and metals) [35]; however, only marginal improvements in efficiency have been achieved, due largely to the rapid charge carrier recombination rate.

Nanocomposites based on carbon allotropes, including multi-walled CNTs (MWCNTs) and RGO, have recently received much attention, owing to their marked electrical and thermal conductivities, large specific surface area, stability in both acidic and basic solvents, and high degree of crystallinity. These materials have been used separately, as well as in conjunction with semiconductor nanostructures, and have been shown to be effective in applications including photocatalysis, supercapacitors, Li-ion batteries, solar cells, and chemical sensors [36]. However, these devices have not exhibited particularly favorable performance because of agglomeration and the limited control over the alignment-dependent properties of the MWCNTs and RGO sheets. The optical properties of RGO depend strongly on the number of layers, and $\pi-\pi$ interaction between adjacent sheets may lead to aggregation [37]; hence, RGO sheets alone may not be sufficient to result in high-performance photocatalysts. MWCNTs, however, have an electrical conductivity of $\sim >10^5$ cm^2 V^{-1} s^{-1} comparable to that of RGO, which is $\sim 2 \times 10^5$ cm^2 V^{-1} s^{-1}. Therefore, MWCNTs can inhibit the aggregation of the RGO sheets, leading to a large specific surface area [38]. These properties indicate that MWCNT/RGO composites can enhance the performance of devices significantly. In particular, their use in photocatalysis has received much attention due to the effective transfer of electrons, which can slow the rate of electron–hole recombination. These composites also exhibited a large specific surface area with a large density of active centers.

In this section, a composite material consisting of functionalized MWCNTs, RGO sheets, and Fe$_2$O$_3$ nanorhombohedra was fabricated using a facile hydrothermal method. The photocatalytic performance of these composites was studied toward degradation of Rhodamine B (RhB) under visible irradiation, using samples of Fe$_2$O$_3$ nanorhombohedra and composites with MWCNTs and RGO.

3.4.1 Preparation of Photocatalyst Powder

Single crystalline hematite nanorhombohedra were synthesized using the hydrothermal method using a 50:50 vol.% mixture of distilled water. Ferric chloride (4.05 g; FeCl$_3$:6H$_2$O) and HMTA (2.10 g) were added with constant stirring for 30 min at room temperature, forming a transparent deep red solution. This solution was transferred into a Teflon-lined stainless steel autoclave with a volume of 100 mL and heated to 150°C for 5 h. The autoclave was then allowed to cool to room temperature naturally and the precipitated solution was filtered. The resulting deep red powders were washed with ethanol and distilled water several times and used in further measurements. To fabricate composite samples, 20 mg of RGO powder was dispersed in 60 ml of a water and ethanol mixture (50:50 vol.%). The resulting solution was ultrasonicated for 1 h to obtain a uniform dispersion of RGO sheets. Identical quantities of $FeCl_3:6H_2O$ and HMTA as described above were dissolved in the RGO dispersion and stirred for 30 min at room temperature. The mixture was transferred to an autoclave and heated at 150°C for 5 h, and then allowed to cool naturally and filtered at room temperature.

One hundred milligrams of MWCNTs powder were added to 250 ml of concentrated HNO_3 and refluxed at 90°C for 2 h. Following this oxidation process, the MWCNT solution was filtered and washed with distilled water to remove the acid residue. The filtrate was dried in an autoclave at 60°C, and the resulting powder samples were collected. To form the $MWCNT/Fe_2O_3$ composite, 20 mg of MWCNT powders was dispersed in a water and ethanol mixture by sonication. Following the formation of a uniform dispersion of MWCNTs, optimal quantities of $FeCl_3:6H_2O$ and HMTA were added and dissolved, and the solution was transferred to an autoclave for the hydrothermal process. The $MWCNT/RGO/Fe_2O_3$ composites were prepared as follows: Powder samples of MWCNTs (20 mg) and RGO (20 mg) were dispersed using ultrasonication in a 50:50 wt% ethanol/water mixture, and an optimized mass of $FeCl_3:6H_2O$ and HMTA was added to the solution. After 30 min of stirring at room temperature, the solution was transferred to an autoclave and heated to 150°C for 5 h. The resulting composite precipitate was filtered, washed with ethanol and water, and dried at 60°C. The obtained powder samples were used for characterization and photocatalysis measurements.

3.4.2 Analysis of MWCNT/RGO/Fe₂O₃ Composites

XRD patterns of powder samples of Fe_2O_3, $Fe_2O_3/MWCNT$ composite, Fe_2O_3/RGO composite, and $Fe_2O_3/MWCNT/RGO$ composite are shown in Figure 3.16. The diffraction peaks for Fe_2O_3 and the composite samples were indexed using a single crystalline rhombohedral phase of hematite (JCPDS#01-087-1164). The sharp peaks in the patterns indicate highly crystalline structure of the hematite, and no peaks related to other phases of iron oxide were found in the diffraction patterns, indicating a pure hematite phase. For the $Fe_2O_3/MWCNT$, Fe_2O_3/RGO, and $Fe_2O_3/MWCNT/RGO$ composite samples, the patterns represent the peaks related to the hematite structure and that of RGO, and peaks corresponding to the MWCNTs, were absent.

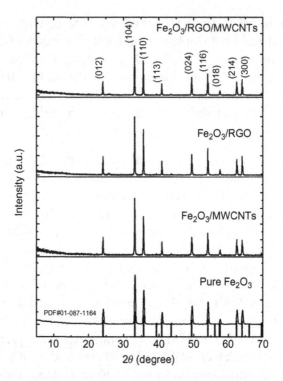

Figure 3.16 XRD patterns for pure hematite and its composites with MWCNTs and RGO sheets. The standard pattern of JCPDS # 01-087-1164 is shown for comparison.

This may be attributed to complete coverage of the RGO sheets with Fe_2O_3 nanorhombohedra or particles, together with good dispersion of the MWCNTs throughout the composite sample. Moreover, there were no noticeable shifts in the peaks in the diffraction pattern. These results indicate that the Fe_2O_3 nanorhombohedra were well distributed across the surface of the RGO sheets and combined completely with the MWCNTs during the hydrothermal process.

As shown in Figure 3.17a, nanoscale rhombohedra were formed, which were highly uniform and exhibited well-defined shapes throughout the entire area of the sample. The magnified image in the inset shows that the nanorhombohedra were a few tens of nanometers in length and in width. During hydrolysis of $FeCl_3$, it is first converted into a hydroxide, and then into an oxide following calcination. Here, we used HMTA as a source of hydroxide (OH^-) ions at elevated temperature to achieve hydrolysis of $FeCl_3$. Furthermore, the HMTA can

Figure 3.17 FESEM images of nanorhombohedral structured hematite and its composites at a magnification of ×25,000: (a) Fe₂O₃, (b) Fe₂O₃/MWCNT composite, (c) Fe₂O₃/RGO composite, and (d) Fe₂O₃/MWCNT/ RGO composite. The insets show the corresponding images at a magnification of × 100,000.

act as shape-inducing molecules, due to the existence of the amine group [39]. The reactions occurring during the growth of hematite nanorhombohedra are as follows:

$$(CH_2N_4) + 6H_2O \rightarrow 6HCHO + 4NH_3 \qquad (3.16)$$

$$NH_3 + H_2O \rightarrow NH_4^+ + OH^- \qquad (3.17)$$

$$FeCl_3 + 3NH_3 \cdot H_2O \rightarrow FeOOH + 3NH_4Cl + H_2O \qquad (3.18)$$

and

$$2FeOOH \rightarrow Fe_2O_3 + H_2O \qquad (3.19)$$

From Eqs. (3.16)–(3.19), we can see that the hydrolysis of $FeCl_3$ occurs, and that the Fe_2O_3 product can be expected to have a rhombohe-dral structure. Hence, HMTA provides OH^- ions for the formation of hematite, which then adsorb onto a particular surface; this minimizes the growth in that direction, resulting in the formation of nanorhombohedral hematite. Following the synthesis of the Fe_2O_3 nanorhombohedra, we combined this structure with functionalized MWCNTs and RGO sheets using ultrasonication followed by a hydrothermal process.

Figure 3.17b shows FESEM images of a Fe_2O_3/MWCNT sample. We find that Fe_2O_3 nanorhombohedra were wrapped with MWCNTs. The low-magnification image reveals a uniform distribution of MWCNTs inside the whole area of the sample. However, the high-magnification image shows that only a small percentage of the MWCNTs were agglomerated (see the inset of Figure 3.17b). For the Fe_2O_3/RGO sample, we observed that micron-sized RGO sheets were covered with Fe_2O_3 nanorhombohedra. The RGO sheets were large compared with the rhombohedra, hence some of the RGO sheets were not covered, as shown in Figure 3.17c. The high-magnification image shows a uniform distribution of rhombohedra over the RGO sheets (see the inset of Figure 3.17c). The Fe_2O_3/MWCNT/RGO sample exhibited MWCNTs, as well as nanorhombohedra, which were combined and distributed across the surface of the RGO sheets, as shown in Figure 3.17d. The high-magnification image shows that nanoscale MWCNTs were wrapped into nanorhombohedra and RGO sheets (see the inset of Figure 3.17d). This formation of heterojunctions is expected to reduce the charge carrier recombination rate compared with Fe_2O_3.

To further investigate the crystal structure of the Fe_2O_3 nanorhombohedra, as well as the composites with MWCNTs and RGO sheets, samples were analyzed using TEM. The Fe_2O_3 powder samples exhibited nanorhombohedra with faceted surfaces, which had an average width of 70 nm and length of 100 nm, as shown in Figure 3.18a. The selected-area electron diffraction (SAED) patterns shown in the inset to Figure 3.18a reveal that the Fe_2O_3 was crystalline. The Fe_2O_3/MWCNTs samples exhibited nanorhombohedra with similar dimensions to the Fe_2O_3 enclosed within MWCNTs, as shown in Figure 3.18b, in which the inset shows the SAED patterns, which indicate the crystalline structure of hematite. This shows that the presence of the MWCNTs did not affect the crystallinity of the hematite. The Fe_2O_3/RGO sample exhibited a distribution of nanorhombohedra on the surface of RGO flakes, as shown in Figure 3.18c; however, the nanorhombohedra did not completely cover the RGO sheet. The inset shows SAED patterns; the crystalline nature of these nanorhombohedra

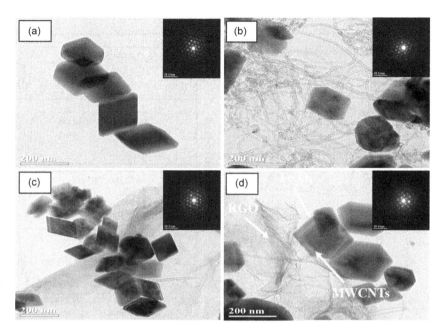

Figure 3.18 TEM images showing the detailed structure of the nanorhombohedral structure of hematite and its composites. (a) Fe_2O_3, (b) Fe_2O_3/MWCNT composite, (c) Fe_2O_3/RGO composite, and (d) Fe_2O_3/MWCNT/ RGO composite. The inset shows the corresponding selected area electron diffraction (SAED) pattern, indicating the single crystalline structure of the hematite rhombohedra.

indicates that RGO had no significant influence on the crystallinity of the hematite structure. Figure 3.18d shows a TEM image of the Fe_2O_3/ MWCNT/RGO sample, which reveals that the nanorhombohedra and MWCNTs covered the surface of RGO sheet, forming ternary heterojunctions. Furthermore, the SAED patterns shown in the inset reveal the single crystal nature of the hematite nanorhombohedra. Therefore, the RGO and MWCNTs do not significantly affect the crystallinity of the hematite. The TEM analyses show that the nanorhombohedra were single-crystal structures that were a few tens of nanometers in size, which is consistent with the XRD patterns shown in Figure 3.16.

FTIR spectra were measured to investigate the functionalization of the MWCNTs, and the formation of the composites with RGO and Fe_2O_3. Initially, we combined these MWCNTs with RGO and Fe_2O_3 structures using the hydrothermal method to form the composite samples. Figure 3.19 shows FTIR spectra of Fe_2O_3, and the Fe_2O_3/ MWCNTs, and the Fe_2O_3/RGO and Fe_2O_3/MWCNT/RGO composite samples. The two peaks at 479 and 559 cm^{-1} correspond to

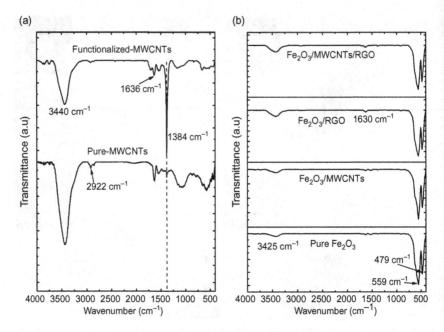

Figure 3.19 FTIR spectra of (a) MWCNTs and functionalized MWCNTs in an acid bath of HNO₃, and (b) Fe₂O₃, as well as the Fe₂O₃/MWCNT, Fe₂O₃/RGO, and Fe₂O₃/MWCNT/RGO composite samples.

vibrational modes of the Fe_2O_3 nanostructure [40]. In addition, there were two minor bands at 3425 and $1636 \, cm^{-1}$, which were attributed to hydrogen-bonded O−H moieties, and are characteristic stretching modes of C−H groups. These groups may have been introduced during the acid treatment, or the synthesis of the MWCNTs. The strong peak related to −NO₂ disappeared due to the formation of bonds between the RGO and Fe_2O_3. The FTIR spectra therefore indicate that composite structures of MWCNTs, RGO, and Fe_2O_3 were successfully formed.

3.4.3 Photocatalytic Performance

To demonstrate the potential of these composite materials for photocatalytic degradation of environmental pollutants, we investigated the photocatalytic degradation of RhB under visible irradiation. The photocatalytic activity of the Fe_2O_3, Fe_2O_3/MWCNT, Fe_2O_3/RGO, and Fe_2O_3/MWCNT/RGO samples was evaluated from the degradation of RhB dye for 2 h, as given in Figure 3.20a. The degradation of the RhB solution under identical experimental conditions but with no photocatalyst is provided for comparison. We can see that the Fe_2O_3/

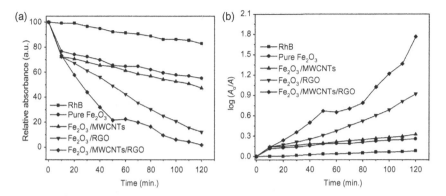

Figure 3.20 (a) The relative RhB concentration during photocatalysis, in the presence of Fe$_2$O$_3$, as well as the Fe$_2$O$_3$/MWCNT, Fe$_2$O$_3$/RGO, and Fe$_2$O$_3$/MWCNT/RGO composite samples. The absorbance of RhB without photocatalysis is also provided for comparison. (b) Logarithmic plots, i.e., ln(A$_o$/A(t)) for RhB degradation with and without the catalysts as a function of time.

Table 3.4 The Specific Surface Area, Reaction Rate Constants (min^{-1}), TOC of RhB (μg C l^{-1}), and Current Density of the Fe$_2$O$_3$, Fe$_2$O$_3$/MWCNT, Fe$_2$O$_3$/RGO, and Fe$_2$O$_3$/MWCNT/RGO Samples

Sample Name	Surface Area (m^{2}g^{-1})	Kinetic Rate Constant (min^{-1}) × 10^{-3}	Total Organic Carbon (TOC) of RhB (μgC l^{-1})	Current Density (mm cm^{-2})
RhB	–	0.07	39670	–
Pure Fe$_2$O$_3$	5.54	1.65	29828	14
Fe$_2$O$_3$/ MWCNTs	10.81	2.03	5081	34
Fe$_2$O$_3$/RGO	12.48	7.01	4730	98
Fe$_2$O$_3$/ MWCNTs/ RGO	30.64	12.79	2535	495

MWCNT/RGO composite sample degraded to 98% of RhB after 2 h of irradiation with visible light, compared with 88% for the Fe$_2$O$_3$/ RGO composite, 53% for the Fe$_2$O$_3$/MWCNT composite, and 45% for Fe$_2$O$_3$. These results clearly show that the Fe$_2$O$_3$/MWCNT/RGO sample exhibited improved photocatalytic activity.

The photocatalytic kinetics of the samples were analyzed using the Langmuir–Hinshelwood model, as shown in Figure 3.20b. All of the data follow a first-order reaction model, and the calculated apparent kinetic rate constants are summarized in Table 3.4. We find that the apparent reaction rate constant for the Fe$_2$O$_3$/MWCNT/RGO sample is $k = 12.79 \, \text{min}^{-1}$, which is approaching sevenfold that of the Fe_2O_3 sample ($k = 1.65 \, \text{min}^{-1}$). The rate constant for Fe_2O_3/RGO is $k = 7.01 \, \text{min}^{-1}$, and that for Fe_2O_3/MWCNTs is $k = 2.03 \, \text{min}^{-1}$, which is also larger than that of pure Fe_2O_3 but smaller than that of the ternary composite. To confirm complete degradation of RhB molecules, we measured TOC of dye solution after photocatalysis experiment. As provided in Table 3.4, the TOC of degraded RhB solution significantly decreased compared with that of without catalyst. Moreover, the lowest TOC amount was found for Fe_2O_3/MWCNT/RGO sample, which supports the kinetic rate constant calculation. Therefore, TOC analysis confirms the photodegradation of RhB under visible irradiation using Fe_2O_3/MWCNT/RGO composites.

Photocatalytic activity depends on effective separation and transport of electron—hole pairs following the absorption of light, the number of available active sites for adsorption of the compound to be catalyzed, and the optical absorbance. The improved photocatalytic activity of the composite materials may be attributed in part to the large specific surface area, which results in a greater density of catalytic centers, and increase in optical absorbance at visible wavelengths. This increases the number of photogenerated electron—hole pairs. However, the highly conducting MWCNTs and RGO sheets capture photoelectrons effectively, which are subsequently transferred toward the pollutants for catalytic degradation; the addition of such conductive carbon allotropes has been shown to increase the photocatalytic performance substantially [41]. Moreover, the presence of the MWCNTs minimizes the spatial gap between the RGO sheets and hematite nanorhombohedra, resulting in favorable transport properties of photoelectrons, and hence improved photocatalytic activity.

3.4.4 Photocatalysis Mechanism

Hematite is a promising candidate for photocatalysis applications; however, its use in photocatalysis has been limited by the high charge carrier recombination rate. We have explored hematite heterojunctions with MWCNTs and RGO to improve the photocatalytic activity. The substantial enhancement in photocatalytic performance may be attributed to effective transport of photoelectrons from the hematite into RGO and/or MWCNTs. Figure 3.21 shows possible pathways for the transfer of photoelectrons involved in the degradation of RhB.

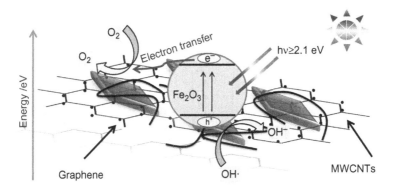

Figure 3.21 The electron transfer mechanism of the Fe$_2$O$_3$/MWCNT/RGO composite under irradiation with visible light.

The work function of the materials is a crucial parameter affecting the transfer of electrons and holes, and rapid transport and separation of charges is required for effective photocatalytic activity. The stepwise energy levels of MWCNTs (-4.8 eV vs. vacuum), RGO (-4.42 eV) and the hematite nanostructures (-4.1 eV) make the composite suitable for the separation of electrons and holes following irradiation [42,43]. Possible electron-transfer mechanisms following illumination with visible light are as follows:

$$Fe_2O_3 + RGO + MWCNTs \xrightarrow{h\nu} Fe_2O_3(h_{VB}^+) + RGO(e^-) \\ + MWCNTs(e^-) \tag{3.20}$$

$$MWCNTs(e^-) + O_2 \rightarrow MWCNTs + O_2^- \tag{3.21}$$

$$Fe_2O_3(h_{VB}^+) + H_2O \rightarrow Fe_2O_3 + OH^\bullet \tag{3.22}$$

Following irradiation, electrons are excited into the conduction band of the hematite, creating holes in the valence band. The photo-generated electrons may transfer into the MWCNTs through the RGO sheets, which reduces the electron–hole recombination rate (Eq. 3.20). The electrons that accumulate in the MWCNTs react with oxygen to form transient superoxide radicals and superoxide molecules (Eqs. 3.21 and 3.6). Simultaneously, the holes in the Fe$_2$O$_3$ (h$_{VB}^+$) form highly reactive OH$\cdot$ radicals (Eq. 3.22), which degrade the RhB dye, as shown in Eqs. 3.7 and 3.8. The incorporation of MWCNTs and RGO in the composite material reduces the charge carrier recombination rate and thereby improves the photocatalytic activity.

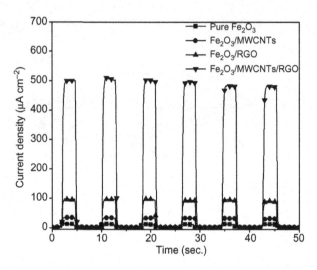

Figure 3.22 PEC measurements showing the photocurrent in Fe₂O₃, as well as the Fe₂O₃/MWCNT, and Fe₂O₃/ RGO and Fe₂O₃/MWCNT/RGO composite samples, as a function of time and at different on- and off-cycles under irradiation with visible light in a 0.1 M NaOH electrolyte.

3.4.5 PEC Measurements

To determine the influence of the MWCNTs and RGO on the recombination of photogenerated electron–hole pairs, we measured the photocurrent response of Fe_2O_3 and the composite samples during several on–off cycles of visible irradiation. A rapid response was observed during each on–off cycle, whereby the photocurrent was high when the sample was illuminated, and approached zero when the irradiation was switched off. From Figure 3.22, we can see that the $Fe_2O_3/$ MWCNT/RGO composite exhibited the highest photocurrent density of $\sim 495\ \mu A/cm^2$, which is almost 35-fold greater than that observed in the Fe_2O_3 sample ($14\ \mu A\ cm^{-2}$). The other two composites exhibited photocurrent densities of $98\ \mu A\ cm^{-2}$ (Fe_2O_3/RGO) and $34\ \mu A\ cm^{-2}$ ($Fe_2O_3/MWCNT$), which are higher than that of the hematite sample, but lower than that of the $Fe_2O_3/MWCNT/RGO$ composite sample. From these data, it is apparent that the formation of composites with RGO and MWCNTs improved the electron transport properties and reduced the charge carrier recombination rate. The inclusion of these two conducting materials with the Fe_2O_3 nanorhombohedra further improved the charge transport properties. These photoelectrochemical measurements show that the $Fe_2O_3/MWCNT/RGO$ composites are an efficient charge transport material, leading to favorable photocatalytic activity.

3.4.6 Nano-heterojunction of g-C₃N₄/CdS/RGO

Since the water splitting was discovered by Fujishima and Honda, semiconductor photocatalysis technology has attracted considerable attention because of the potential applications in photodegradation of organic pollutants, self-cleaning, and bacterial elimination using non-toxic, low-cost methods [44]. This represents one of the most promising technologies to overcome the present energy and water pollution problems. Over the last four decades, various semiconductor photocatalysts, including TiO_2, SnO_2, ZnO, WO_3, Fe_2O_3, Bi_2WO_6, In_2S_3, and BiOBr, have been studied and employed in environmental remediation [45,46]. Despite the high photocatalytic activity and good chemical stability, the application of semiconductor catalysts is limited because of the high rate of recombination of electron–hole pairs, low absorption coefficient, and mismatch with the solar spectrum. Therefore, various approaches, including semiconductor–semiconductor, elemental doping in semiconductor, semiconductor–metal, semiconductor–graphene, and sensitization with visible band gap semiconductors, have been developed [47]. The development of semiconductors that absorb in the visible part of the electromagnetic spectrum and the use of graphene are among the most promising strategies due to the large coverage of the visible region of solar spectrum (40%) and reduced recombination of photogenerated charge carriers. Recently, a polymeric semiconductor, known as graphitic carbon nitride (g-C₃N₄), has been reported with remarkable potential applications in photocatalysis and hydrogen gas production [48]. It exhibits various interesting properties, including visible-light absorption (with a band gap of 2.7 eV), thermal, chemical (acidic and basic solutions), and photochemical stability due to its s-triazine structure, and a high degree of condensation. Moreover, the method of fabrication is simple and low cost (i.e., sintering of nitrogen-rich compounds such as urea, thiourea, and melamine) [49]. Additionally, the band structure may be tailored by modification in the nanostructure and by chemical doping.

Nevertheless, bare g-C₃N₄ suffers from a high recombination rate of the photogenerated electron–hole pairs, resulting in poor photocatalytic performance. Various strategies have been reported to overcome these problems [50]. However, these routes exhibit rapid recombination and low stability, resulting in poor photocatalytic performance. Conducting materials, including RGO, CNTs, and gold (Au) nanoparticles have been combined with g-C₃N₄ to resolve these problems [51].

Of these, RGO appears the most promising material for effective transfer of photoelectrons due to the high electron mobility at room temperature, large specific surface area, excellent thermal conductivity, and high Young's modulus [52]. RGO has been explored by combining with various semiconductors to develop efficient photocatalysts [53]. It was observed that RGO played a crucial role in the effective separation of photogenerated electron–hole pairs. Despite RGO incorporation, the photocatalytic performance of g-C_3N_4 has not been improved to an extent that it is suitable for industrial or commercial use. Recently, sensitization with visible band gap semiconductors, including CdS, CdSe, and PbS, has been shown to be a promising route to enhance photodegradation [54]. From the above discussion of strategies, it appears that combining multiple materials allows us to exploit the various material properties, and composite materials may be expected to be particularly suitable for the development of efficient photocatalysts. Thus, a combination of CdS and RGO together with g-C_3N_4 photocatalyst may show superior photocatalytic activity. In a preliminary study, we found that CdS and RGO improved the photodegradation performance of ZnO nanorods due to increasing visible absorption and separation of the electron–hole pairs [37,38]. In this section, we will study g-C_3N_4/CdS/RGO composites fabrication using a facile aqueous chemical method for the photodegradation of RhB and Congo red (CR) dyes under visible irradiation.

3.4.7 Preparation of g-C_3N_4 Powder

To collect g-C_3N_4 powders, the authors followed the method reported in Ref. [55]. In brief, 20 g of urea was placed in a ceramic crucible and placed in an oven at 70°C for 2 hours. The crucible was then placed in a box furnace and annealed at 580°C under air atmosphere for 2 hours with its heating rate at 5°C per minute. After the furnace was allowed to cool to room temperature, the yellow g-C_3N_4 powders were collected in glass vials.

The composites were fabricated using a CBD method. Initially, a 100 mg of g-C_3N_4 powders was dispersed in 80 ml of distilled water by ultrasonication for 1 hour. To fabricate the g-C_3N_4/CdS composite, an equimolar quantity of a 0.005 M solution of cadmium sulfate (0.153 g) and thiourea (0.017 g) was added to the g-C_3N_4 dispersion, which was stirred constantly, and the pH of solution was adjusted to 11 by adding ammonia. The solution was heated at 70°C for 5 hours while stirring constantly. g-C_3N_4/CdS powders were then collected by filtration, and washed with distilled water and ethanol. Similar procedure was used to collect RGO/CdS powders. To fabricate the g-C$_3$N$_4$/RGO composite, 0.20 wt% RGO (with respect to weight of g-C$_3$N$_4$) and 100 mg of g-C$_3$N$_4$ were mixed and dispersed in 80 ml of distilled water using ultrasonication for 1 hour. The solution was then filtered, and powdered g-C$_3$N$_4$/RGO was collected.

To fabricate the g-C$_3$N$_4$/CdS/RGO composite, 0.20 wt% RGO and a 100 mg of g-C$_3$N$_4$ was dispersed in 80 ml of distilled water using ultrasonication, and cadmium sulfate and thiourea were added to the mixture under constant stirring at room temperature. The pH of the mixture was maintained at 11 by adding ammonia, and it was heated at 70°C for 5 hours. The powdered g-C$_3$N$_4$/CdS/RGO composite was then collected by filtration.

3.4.8 Analysis of g-C$_3$N$_4$/CdS, g-C$_3$N$_4$/RGO, and g-C$_3$N$_4$/CdS/RGO Composites

Formation of porous and sheet-like structure of pure g-C$_3$N$_4$ sample was studied using FESEM (see Figure 3.23a). It can be seen that the obtained pure g-C$_3$N$_4$ structure is made up of few nanometer sized sheets and pores. Further, from the magnified FESEM image, it revealed that the g-C$_3$N$_4$ sheets were interconnected and small pores were formed in between two adjacent sheets. In g-C$_3$N$_4$/CdS composite, the aggregation of CdS particles with a random shape and size were obtained, and these were combined throughout the sample with g-C$_3$N$_4$ sheet (see Figure 3.23b). Therefore, it was difficult to differentiate between these two materials (see the magnified image of Figure 3.23b). For the g-C$_3$N$_4$/RGO composite, it can be seen that smaller g-C$_3$N$_4$ sheets were formed as these were coated on the top surface of RGO, as shown in Figure 3.23c. The magnified image of g-C$_3$N$_4$/RGO composite indicates that the RGO sheets mixed well with the g-C$_3$N$_4$ sheets, which were several nanometers thick (see the magnified image of Figure 3.23c). In case of g-C$_3$N$_4$/CdS/RGO composite, it can be seen that g-C$_3$N$_4$ sheets formed right at the surface of the RGO, which was distributed uniformly throughout sample (see Figure 3.23d). Following incorporation of the CdS nanoparticles (CNPs), most of the g-C$_3$N$_4$ sheets were covered with the particles, which had an average diameter of 40 nm (see the magnified image of Figure 3.23d). The density of CNPs was high, as was that of the g-C$_3$N$_4$ sheets, and the CNPs were distributed over the entire sample

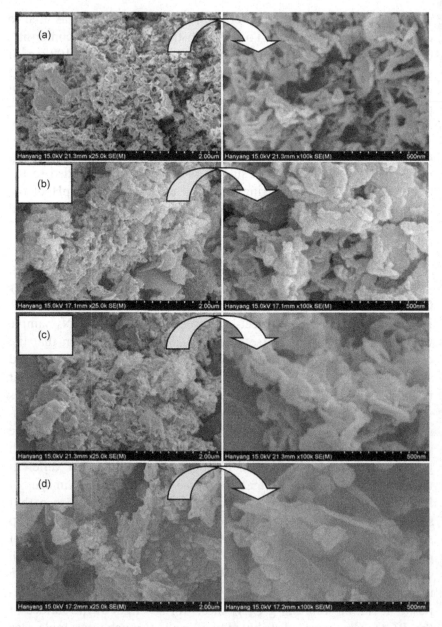

Figure 3.23 FE-SEM images. (a) g-C₃N₄ sheets, (b) The g-C₃N₄ and CdS (0.005 M) combined using a chemical route, (c) The g-C3N4/RGO composite (0.20 wt%), and (d) The g-C₃N₄/CdS/RGO composite at magnifications of 25,000× and 100,000×.

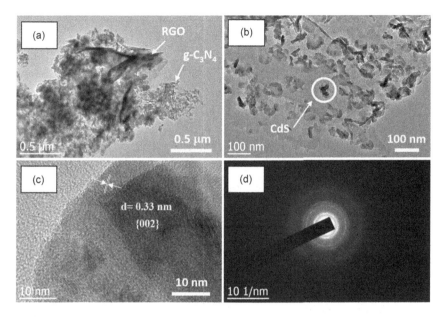

Figure 3.24 (a) TEM image of the g-C₃N₄/CdS/RGO composite with 0.005 M CdS and 0.20 wt% RGO sheets. (b) A magnified TEM image showing the CdS nanoparticles on the g-C₃N₄ sheets, (c) HR-TEM image of CdS particles showing interplanar spacing of 0.33 nm oriented along (002) direction, and (d) SAED pattern of corresponding area revealed crystalline structure of CdS.

with no evidence of aggregation. This indicates that the presence of the RGO sheets may have hindered the aggregation of the CNPs. We have previously analyzed RGO and confirmed the formation of RGO sheets with an extent of several micrometers [52].

Then, the TEM images of g-C₃N₄ and CdS nanoparticles combined with the RGO sheets to form the g-C₃N₄/CdS/RGO composite, are shown in Figure 3.24. It can be seen that the CdS nanoparticles and g-C₃N₄ sheets completely covered the RGO sheets; however, the CdS nanoparticles appeared invisible. The presence of micron-sized RGO sheets also confirms that the RGO sheets were largely intact (Figure 3.24a). The magnified TEM image in Figure 3.24b shows that the CdS nanoparticles were sporadically coated on the g-C₃N₄ and RGO sheets. To confirm the presence of the CdS nanoparticles, we further analyzed the g-C₃N₄/CdS/RGO composite using HR-TEM. Figure 3.24c shows that the HRTEM image of CNPs were distributed over the entire g-C₃N₄ and RGO sheets. The measured lattice spacing for crystalline plane of 0.33 nm indicated the CdS particles with a hexagonal crystal structure oriented along (002) direction (JCPDS#01-070-2553). The nano-crystalline nature of CdS particles was confirmed

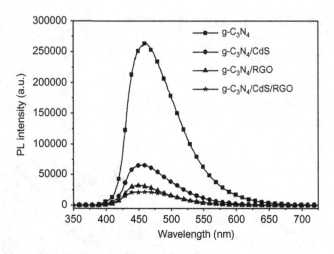

Figure 3.25 Room temperature photoluminescence spectra recorded in the range 350–750 nm with an excitation wavelength of 325 nm for pure g-C₃N₄ and the g-C₃N₄/CdS, g-C₃N₄/RGO, and g-C₃N₄/CdS/RGO composites.

from selected-area electron diffraction (SAED) pattern as shown in Figure 3.24d, and it matches with the pattern for pure CdS nanoparticles. The presence of the CdS nanoparticles, g-C_3N_4, and RGO sheets in TEM images, confirms the formation of the g-C_3N_4/CdS/RGO composite. Moreover, the high density of the g-C_3N_4 sheets and CNPs, is expected to enhance optical absorption because of its band-gap energies in visible wavelength, and RGO is expected to minimize recombination losses by allowing separation of the electron−hole pairs.

Effective charge separation and transport can be determined from room temperature photoluminescence (PL) spectra. Figure 3.25 shows PL spectra of pure g-C_3N_4, g-C_3N_4/CdS, g-C_3N_4/RGO, and g-C_3N_4/CdS/RGO samples measured at an excitation wavelength of 325 nm. It can be seen that the pure g-C_3N_4 and its composites exhibit similar emission trends, with the principle peak around 450 nm, which is attributed to n-π* electronic transitions [56]. The energy of this peak was approximately equal to the band-gap energy of g-C_3N_4 (i.e., 2.7 eV). The pure g-C_3N_4 sample was found to have a higher PL intensity than the composites. This indicates that g-C_3N_4 had the highest optical recombination rate, which may deteriorate the photodegradation efficiency. However, PL quenching was observed in the composites, and it was most significant in g-C_3N_4/CdS/RGO, which indicates that this composite had a lower rate of recombination of photoelectrons compared with pure g-C_3N_4 and the other composites. The low emission of g-C$_3$N$_4$/CdS/RGO composite is attributed to the electron transfer capability of the RGO sheets because of high charge-carrier mobility. Hence, we believe that the g-C$_3$N$_4$/CdS/RGO composite will degrade organic pollutants more rapidly than will either the pure g-C$_3$N$_4$ or the other composites.

3.4.9 Photocatalytic Degradation of RhB Under Visible Irradiation

The photocatalytic activity of pure g-C$_3$N$_4$, as well as the composites with CdS and RGO, was assessed by degrading RhB and Congo red (CR) solutions under irradiation with visible light. Moreover, dark adsorption was measured for 30 minutes of dye molecules to check self-degradation. As a result, the optical absorbance remained almost the same for pure and composite samples indicating that there is no adsorption in the dark. The photocatalytic activities of the pure and composite samples were analyzed by plotting the relative absorbance 100 $A_o/A(t)$, as a function of the irradiation time (Figure 3.26a).

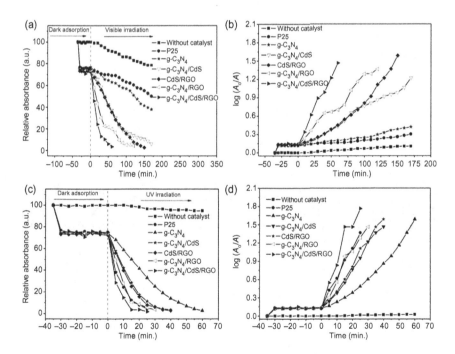

Figure 3.26 (a) Relative optical absorbance and (b) ln(A_o/A(t)) for P25, pure g-C$_3$N$_4$, and g-C$_3$N$_4$/CdS, RGO/ CdS, g-C$_3$N$_4$/RGO, and g-C$_3$N$_4$/CdS/RGO composites as a function of visible irradiation time. The absorbance of the RhB solution under visible-light irradiation without a catalyst is shown for comparison: (c) Relative optical absorbance and (d) ln(A_o/A(t)) for RhB photodegradation without catalyst and with P25, pure g-C$_3$N$_4$, g-C$_3$N$_4$/ CdS, RGO/CdS, g-C$_3$N$_4$/RGO, and g-C$_3$N$_4$/CdS/RGO as a function of UV irradiation time.

Photodegradation of RhB without any catalyst and with Degussa P25 powder (TiO_2) was used as a control. With no catalyst, a slight decrease in the RhB concentration was observed after the visible-light irradiation. However, the RhB concentration decreased considerably in the presence of the catalysts. The g-C_3N_4/CdS/RGO composite exhibited superior photocatalytic activity to the other composites, as well as to pure g-C_3N_4. The improved photocatalytic activity of the g-C_3N_4/CdS/RGO composite is attributed to reduced recombination losses compared with pure g-C_3N_4.

We have analyzed the photodegradation properties of the materials by calculating the kinetic rate constant from $\ln(A_o/A(t))$ versus t, where k is the gradient of this curve, that is, the apparent kinetic rate constant. The linear curve shown in Figure 3.26b illustrates that first-order reaction kinetics apply, which follow the Langmuir–Hinshelwood model. The kinetic rate constant for g-C_3N_4/CdS/RGO was found to be $k = 21.92 \text{ min}^{-1}$, which is almost twenty times larger than that of P25 ($k = 0.96 \text{ min}^{-1}$) and was the highest among all g-C_3N_4-based samples. The rate constants of the pure g-C_3N_4 ($k = 1.77 \text{ min}^{-1}$), g-$C_3N_4$/CdS ($k = 6.43 \text{ min}^{-1}$), RGO/CdS ($k = 9.31 \text{ min}^{-1}$), and g-$C_3N_4$/RGO ($k = 10.49 \text{ min}^{-1}$) were all larger than that of the P25 powder, but were significantly smaller than that of the g-C_3N_4/CdS/RGO composite (Table 3.5). Further, TOC concentration after photodegradation of dye was measured to confirm the complete degradation of dye molecules (Table 3.6). A similar trend in TOC was found as that in optical absorbance.

Table 3.5 The Calculated Reaction Rate Constants (min^{-1}) for P25, Pure g-C_3N_4, g-C_3N_4/CdS, CdS/RGO, g-C_3N_4/RGO, and g-C_3N_4/CdS/RGO Toward Photodegradation of RhB and CR

Sample Details	Kinetic Rate Constant $(K \times 10^{-3}) \text{ min}^{-1}$ RhB		Kinetic Rate Constant $(K \times 10^{-3}) \text{ min}^{-1}$ CR	
	VIS	UV	VIS	UV
RhB	0.67	0.43	0.02	0.06
P25	0.96	48.98	1.91	45.33
g-C_3N_4	1.77	23.85	3.24	26.02
g-C_3N_4/CdS	6.43	36.09	4.06	28.78
RGO/CdS	9.31	38.55	5.56	32.04
g-C_3N_4/RGO	10.49	46.02	14.80	43.59
g-C_3N_4/CdS/RGO	21.92	66.12	24.37	50.90

Table 3.6 Total Organic Carbon (TOC) Analysis of Dye Solution, P25, Pure g-C₃N₄, g-C₃N₄/CdS, CdS/RGO, g-C₃N₄/RGO, and g-C₃N₄/CdS/RGO after Complete Degradation Under UV and VIS Light Irradiation ($\mu gC\ l^{-1}$) Toward Photodegradation of RhB and CR

Sample Details	TOC of RhB in mg C l^{-1}		TOC of CR in μg C l^{-1}	
	VIS	UV	VIS	UV
Dye solution	22040	22040	25980	25980
P25	14317	1610	16248	1759
g-C₃N₄	11589	1936	14384	3971
g-C₃N₄/CdS	5794	1610	11942	3067
RGO/CdS	3612	1847	6482	2841
g-C₃N₄/RGO	1936	1737	2385	1951
g-C₃N₄/CdS/RGO	1685	1467	1821	1678

TOC for RhB and for CR dyes without photocatalysts is provided for comparison.

It is known that photodegradation performance of catalysts largely depends on the recombination and transport of photogenerated charge carriers, the optical absorbance, and the available surface area for dye adsorption on the surface of the catalyst. We found that the g-C₃N₄/CdS/RGO composite exhibited the largest optical absorbance and specific surface area of the g-C₃N₄-based materials. Due to large electron mobility of the RGO, captured photoelectrons from the CdS and g-C₃N₄ semiconductors may be effectively transported, resulting in reduced electron–hole recombination, which was consistent with the PL quenching observations.

3.4.10 Photocatalytic Degradation of RhB Under UV Irradiation

After visible irradiation, UV photocatalytic activity of the prepared composites under the same conditions was investigated as that for the visible light. As shown in Figure 3.26c, which is an overall comparison of relative absorbance, RhB degradation in the absence of photocatalyst is very slow. The removal efficiency of g-C₃N₄/CdS/RGO composite is higher than those of the other composites and pure g-C₃N₄. Trend of these results is very similar to the cases with visible photocatalytic activity. However, the photocatalytic activity of P25 is high or similar to that of g-C₃N₄/CdS/RGO composite, attributed to its band gap energy (3.2 eV), which absorbs UV light effectively. For the further analysis of these composites under UV light, kinetic rate constant from logarithmic plots was plotted as shown in Figure 3.26d. A summary of the results is provided in Table 3.5.

It was observed that the calculated rate constant of g-C_3N_4/CdS/RGO composite ($k = 66.12\,\text{min}^{-1}$) is high, compared to those of the pure g-C_3N_4 ($k = 23.85\,\text{min}^{-1}$), g-$C_3N_4$/CdS ($k = 36.09\,\text{min}^{-1}$), RGO/CdS ($k = 38.55\,\text{min}^{-1}$), and g-$C_3N_4$/RGO ($k = 46.02\,\text{min}^{-1}$) composites and P25 ($k = 48.98\,\text{min}^{-1}$) powders. It clearly demonstrated that RGO and CdS improved the photocatalytic activity of pure g-C_3N_4 due to its high rate of photoelectrons transfer and specific surface area for dye photodegradation. TOC analyses of photodegraded RhB under UV light are provided in Table 3.6. It was seen that g-C_3N_4/CdS/RGO composite and P25 powders have the lowest amount of carbon concentration remaining compared to those of other composites and pure g-C_3N_4. Therefore, RGO sheet and CdS nanoparticles improved photocatalytic activity of g-C3N4 due to its high absorption and transportation of photogenerated electrons.

3.4.11 Photocatalytic Degradation of CR Under Visible Irradiation

To use g-C_3N_4-based composites in a wide range of photodegradation applications, the photodegradation of CR dye under irradiation with visible light was explored. All measurements were performed under the same conditions as those for RhB experiment. CR photodegradation was analyzed using the relative absorbance and logarithmic plots as shown in Figure 3.27a, b. The photocatalytic activities with no catalyst and with P25 powders are shown for comparison. It is clear that no photodegradation occurred in the absence of a catalyst, and where a catalyst was used, complete degradation of the dye was observed. The kinetic rate constant calculated from logarithmic plots was found to be $k = 24.37\,\text{min}^{-1}$ for g-C_3N_4/CdS/RGO composite, compared to $k = 1.91\,\text{min}^{-1}$ for P25, $k = 3.24\,\text{min}^{-1}$ for pure g-C_3N_4, $k = 4.06\,\text{min}^{-1}$ for the g-C_3N_4/CdS composite, $k = 5.56\,\text{min}^{-1}$ for the RGO/CdS composite, and $k = 14.80\,\text{min}^{-1}$ for the g-C_3N_4/RGO composite. The g-C_3N_4/CdS/RGO composite resulted in a significantly faster photodegradation of the CR dye compared to the other materials. That is consistent with the photocatalytic degradation results for RhB. After visible photodegradation of CR, we analyzed the remaining solution using TOC analyzer as summarized in Table 3.6. It revealed that the CR solution degraded with g-C_3N_4/CdS/RGO composite, resulting in a lowest amount of carbon concentration among all the samples. The trend of these results matches with that for RhB photocatalytic activity under UV and visible light.

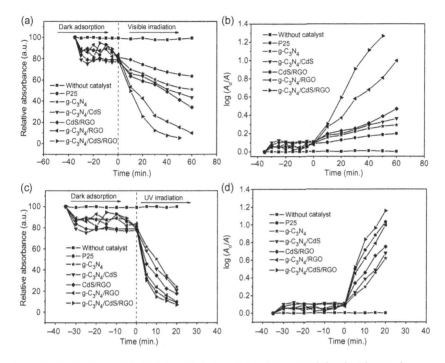

Figure 3.27 (a) Relative optical absorbance and (b) $\ln(A_o/A(t))$ for Congo red photodegradation without catalyst and with P25, pure g-C_3N_4, g-C_3N_4/CdS, RGO/CdS, g-C_3N_4/RGO, and g-C_3N_4/CdS/RGO as a function of visible irradiation time, (c) relative optical absorbance, and (d) $\ln(A_o/A(t))$ for Congo red photodegradation without catalyst and with P25, pure g-C_3N_4, g-C_3N_4/CdS, RGO/CdS, g-C_3N_4/RGO, and g-C_3N_4/CdS/RGO as a function of UV irradiation time.

3.4.12 Photocatalytic Degradation of CR Under UV Irradiation

Finally, we studied the photocatalytic activity of prepared composites under UV irradiation toward CR degradation, and its corresponding UV-VIS optical absorbance plots were analyzed by plotting relative absorbance graphs (Figure 3.27c) and calculating kinetic rate constants from logarithmic plots (Figure 3.27d). The photodegradation of CR with P25 and activity without any catalysts are given for comparison. Apparently, the calculated kinetic rate constant for g-C_3N_4/CdS/RGO composite showed the highest value, that is, $50.90 \, \text{min}^{-1}$, which is almost double than that of pure g-C_3N_4 (Table 3.5). However, the kinetic rate constant for P25 powder ($45.33 \, \text{min}^{-1}$) sample was found to be almost similar with that of g-C_3N_4/CdS/RGO composite, which was attributed to band gap energy of P25 in UV region. The photodegraded CR solution was further analyzed using TOC analyzer and its results are summarized in Table 3.6. As a result, the lowest TOC concentration for g-C_3N_4/CdS/RGO composite revealed that the composition of g-C_3N_4, CdS, and RGO is suitable for superior CR photodegradation, compared to any other combination of composites or pure g-C_3N_4.

3.4.13 Mechanism of Enhanced Photocatalysis

During photocatalysis process, it is known that under irradiation condition, oxygen species, such as superoxide radical ($O_2^{\bullet-}$), hydroxide radical ($OH^{\bullet}$), and hydrogen peroxide (H_2O_2), initiate the photodegradation reactions. These radicals are produced by photocatalytic reduction of oxygen and oxidation of water. The photocatalytic activity of catalyst depends on its ability to reduce and oxidize photoelectrons and holes generated under irradiation. For the present g-C_3N_4/CdS/RGO composite, the photogenerated electrons from g-C_3N_4 (-3.38 eV) can transfer easily through conduction of CdS (-3.98 eV) and then get captured by RGO (-4.4 eV) sheet due to their suitable energy levels [57]. The captured electrons were utilized for reduction of oxygen $(O_2/O_2^{\bullet-}) = -0.16$ eV. Therefore, the conduction band level of g-C_3N_4 is suitable for the generation of superoxide radicals and subsequently OH radicals. However, the valence band level of g-C_3N_4 (1.57 eV) is not sufficient for direct oxidation of H_2O (2.7 eV) molecules through oxidation process. Hence, $OH^{\bullet}$ radicals could be generated from superoxide radicals [58,59]. Figure 3.28 shows possible routes of photoelectron generation and transport in the g-C_3N_4/CdS/RGO composite.

$$\text{g-}C_3N_4 + \text{CdS} + \text{graphene} \xrightarrow{h\nu} \text{g-}C_3N_4(h_{VB}^+) + \text{CdS}(e_{CB}^-) \\ + \text{graphene}(e^-) \tag{3.23}$$

Because of the visible-wavelength band gap energy of g-C_3N_4, it absorbs visible light and generates electron–hole pairs when the photon energy is greater than approximately 2.7 eV (Eq. (3.23)). Photogenerated electrons are then transferred toward the conduction band of CdS and to the RGO sheets (Eq. (3.15)). Then these electrons captured by the RGO sheet react with oxygen to form transient superoxide radicals and superoxide molecules (Eq. (3.6)). Highly reactive -OH radicals form, which mineralize RhB and CR molecules, as shown in Eqs. (3.7) and (3.8). Therefore, CdS and RGO play important roles in minimizing the recombination losses and thereby enhancing the photodegradation.

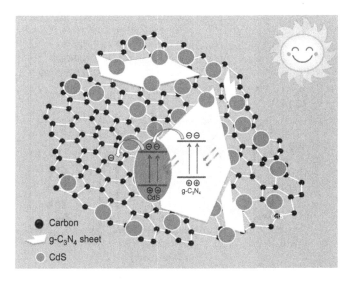

Figure 3.28 The electron transfer mechanism under visible light and the photocatalytic degradation of organic pollutants with the g-C₃N₄/CdS/RGO composite.

The role of CdS nanoparticles and RGO sheets incorporated with the g-C$_3$N$_4$ network was further investigated by measuring the transient photocurrent response of pure g-C$_3$N$_4$, g-C$_3$N$_4$/CdS, g-C$_3$N$_4$/RGO, and g-C$_3$N$_4$/CdS/RGO under irradiation with visible light. The results are shown in Figure 3.29; the photocurrent is rapidly increased following the start of irradiation and returned to zero when the light source was switched off. As shown in the figure, the results exhibited repeatability over the five cycles, indicating that both the pure g-C$_3$N$_4$ and the composites were photochemically stable. It is interesting to note that current density of the g-C$_3$N$_4$/CdS/RGO composite (175 μA cm^{-2}) was almost twelve times larger than that of pure g-C$_3$N$_4$. These results are consistent with the photodegradation data described above.

To determine the photo stability of g-C$_3$N$_4$/CdS/RGO composite under UV and visible light, the cyclic performance was measured up to three cycles (Figure 3.30). It is clearly seen that g-C$_3$N$_4$/CdS/RGO composite exhibits no noticeable change in photodegradation efficiency, although there is infinitesimal decline in performance as compared with that of the first cycle. These results confirmed that g-C$_3$N$_4$/CdS/RGO composite has good stability and could be used in a device for water purification application.

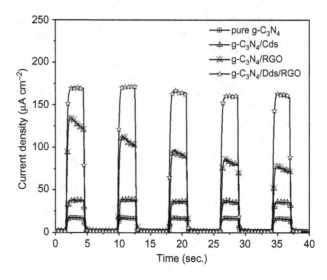

Figure 3.29 Transient photocurrent responses of pure g-C₃N₄, g-C₃N₄/CdS, g-C₃N₄/RGO, and g-C₃N₄/CdS/RGO recorded under excitation with visible light in an 0.5M Na₂SO₄ electrolyte.

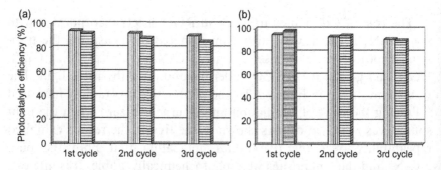

Figure 3.30 Cycling test of g-C₃N₄/CdS/RGO composite up to three times under VIS (bars with vertical lines) and UV (bars with horizontal lines) illumination toward (a) Rhodamine B and (b) Congo red.

3.5 CONCLUSION

In summary, chapter 3 gives a detailed demonstration about different binary (RGO/ZnO, RGO/CdS, RGO/Fe₂O₃, CNTs/Fe₂O₃) and ternary (RGO/CdS/ZnO, RGO/CNTs/Fe₂O₃, and RGO/CdS/g-C₃N₄) photocatalysts fabricated in our laboratory through chemical methods. These photocatalysts were used successfully in water purification application toward degradation of pollutants under visible irradiation. Various analytical tools such as Raman spectroscopy, optical absorbance, X-ray diffraction pattern, Fourier transform infrared spectroscopy, scanning electron microscopy, transmission electron microscopy, and energy dispersion spectroscopy confirmed the formation of above-mentioned heterogeneous nanocomposites. To study effective generation and transportation of photoelectrons, we measured photoelectrochemical properties of these catalysts and found the effective separation of electron–hole pairs. Finally, reusability test confirmed the stability of photocatalysts over five cycles. Therefore, an efficient photocatalyst was fabricated successfully using the facile chemical method. The improved photocatalytic performance resulted in effective separation of photogenerated electro–hole pairs, high specific surface area, and high optical absorbance in visible region. From these studies, we demonstrated that the heterogeneous nanocomposites are applicable in water purification devices.

REFERENCES

[1] Sun JH, Dong SY, Wang YK, Sun SP. Preparation and photocatalytic property of a novel dumbbell-shaped ZnO microcrystal photocatalyst. J Hazard Mater 2009;172:1520–6.

[2] Pawar RC, Shaikh JS, Shinde PS, Patil PS. Dye sensitized solar cells based on zinc oxide bottle brush. Mater Lett 2011;65:2235–7.

[3] Kundu P, Deshpande PA, Madras G, Ravishankar N. Nanoscale ZnO/CdS heterostructures with engineered interfaces for high photocatalytic activity under solar radiation. J Mater Chem 2011;21:4209–16.

[4] Fe Y, Chen H, Sun X, Wang X. Combination of cobalt ferrite and graphene: high-performance and recyclable visible-light photocatalysis. Appl Catal B Environ 2012;111–112:280–7.

[5] Wang J, Tsuzuki T, Tang B, Hou X, Sun L, Wang X. Reduced graphene oxide/ZnO composite: reusable adsorbent for pollutant management. ACS Appl Mater Interfaces 2012;4:3084–90.

[6] Tian C, Zhang Q, Wu A, Jiang M, Liang Z, Jiang B, et al. Cost-effective large- scale synthesis of ZnO photocatalyst with excellent performance for dye photodegradation. Chem Commun 2012;48:2858–60.

[7] Stankovich S, Dikin DA, Piner RD, Kohlhaas KA, Kleinhammes A, Jia Y, et al. Synthesis of graphene-based nanosheets via chemical reduction of exfoliated graphite oxide. Carbon 2007;45:1558–65.

[8] Mohanty N, Nagaraja A, Armesto J, Berry V. High-throughput, ultrafast synthesis of solution-dispersed graphene via a facile hydride chemistry. Small 2010;2:226–31.

[9] Park S, Ruoff RS. Chemical methods for the production of graphenes. Nat Nanotechnol 2009;4:217–24.

[10] Xiang Q, Yu J, Jaroniec M. Graphene-based semiconductor photocatalysts. Chem Soc Rev 2012;41:782–96.

[11] Zhang N, Zhang Y, Yang MQ, Tang ZR, Xu YJ. A critical and benchmark comparison on graphene-, carbon nanotube-, and fullerene-semiconductor nanocomposites as visible light photocatalysts for selective oxidation. J Catal 2013;299:210–21.

[12] Depan D, Shah J, Misra RDK. Controlled release of drug from folate-decorated and graphene mediated drug delivery system: Synthesis, loading efficiency, and drug release response. Mater Sci Eng C 2011;31:1305–12.

[13] Yang L, Xiao Y, Liu S, Li Y, Cai Q, Luo S, et al. Photocatalytic reduction of Cr(VI) on WO$_3$ doped long TiO$_2$ nanotube arrays in the presence of citric acid. Appl Catal B Environ 2010;94:142–9.

[14] Ni Z, Wang Y, Yu T, You Y, Shen Z. Reduction of Fermi velocity in folded graphene observed by resonance Raman spectroscopy. Phys Rev B 2008;77:235403–5.

[15] Zhang D, Wen M, Jiang B, Zhu J, Huo Y, Li G. Microwave-assisted architectural control fabrication of 3D CdS structures. J Sol-Gel Sci Technol 2012;62:140–8.

[16] Pawar RC, Kim H, Lee CS. Improved field emission and photocatalysis properties of cacti-like zinc oxide nanostructures. Scripta Mater 2013;68:142–5.

[17] Liu X, Pan L, Lv T, Zhu G, Sun Z, Sun C. Microwave-assisted synthesis of CdS-reduced graphene oxide composites for photocatalytic reduction of Cr(VI). Chem Commun 2011;47:11984–6.

[18] Yu YJ, Zhao Y, Ryu S, Brus LE, Kim KS, Kim P. Tuning the graphene work function by electric field effect. Nano Letters 2009;9:3430–4.

[19] Bai S, Shen X, Zhong X, Liu Y, Zhu G, Xu X, et al. One-pot solvothermal preparation of magnetic reduced graphene oxide-ferrite for organic dye removal. Carbon 2012;50:2337–46.

[20] Liu X, Pan L, Lv T, Lu T, Zhu G, Sun Z, et al. Microwave-assisted synthesis of ZnO–graphene composite for photocatalytic reduction of Cr(VI). Catal Sci Technol 2011;1:1189–93.

[21] Chakrabarti S, Chaudhuri B, Bhattacharjee S, Das P, Dutta BK. Degradation mechanism and kinetic model for photocatalytic oxidation of PVC–ZnO composite film in presence of a sensitizing dye and UV radiation. J Hazard Mater 2008;154:230–6.

[22] Jang JS, Yu CJ, Choi SH, Ji SM, Kim ES, Lee JS. Topotactic synthesis of mesoporous ZnS and ZnO nanoplates and their photocatalytic activity. J Catal 2008;254:144–55.

[23] Shkrob IA, Sauer MC. Hole scavenging and photo-stimulated recombination of electron – hole pairs in aqueous TiO$_2$ nanoparticles. J Phys Chem B 2004;108:12497–511.

[24] Nozik AJ. Quantum solar cells. Physica E 2002;14:115–20.

[25] Stankovich S, Dikin DA, Dommett GHB, Kohlhaas KM, Zimney EJ, Stach EA, et al. Graphene-based composite materials. Nature 2006;;442:282–6.

[26] Zhou G, Wang DW, Li F, Zhang L, Li N, Wu ZS, et al. Graphene-wrapped Fe$_3$O$_4$ anode material with improved reversible capacity and cyclic stability for lithium ion batteries. Chem Mater 2010;22:5306–13.

[27] Wang Z, Zhang H, Li N, Shi Z, Gu Z, Cao G. The graphene nanosheets shows excellent performance as anode materials of lithium-ion batteries. Nano Res 2010;3:748–56.

[28] Sun Y, Wu Q, Shi G. Graphene based new energy materials. Energy Environ Sci 2011;4:1113–32.

[29] Zeng P, Zhang Q, Peng T, Zhang X. One-pot synthesis of reduced graphene oxide–cadmium sulfide nanocomposite and its photocatalytic hydrogen production. Phys Chem Chem Phys 2011;13:21496–502.

[30] Zedan AF, Sappal S, Moussa S, Shall MS. Ligand-controlled microwave synthesis of cubic and hexagonal CdSe nanocrystals supported on graphene: photoluminescence quenching by graphene. J Phys Chem C 2010;114:19920–7.

[31] Fox M, Dulay M. Heterogeneous photocatalysis. Chem Rev 1993;93:341–7.

[32] Pawar RC, Cho D, Lee CS. Fabrication of nanocomposite photocatalysts from zinc oxide nanostructures and reduced graphene oxide. Curr Appl Phys 2013;13:S50−7.

[33] Sivula K, Formal FL, Gratzel M. Solar water splitting: progress using hematite (α-Fe^{2}O^3) photoelectrodes. ChemSusChem 2011;4:432−49.

[34] Baltrusaitis J, Hu Y-S, McFarland EW, Hellman A. Photoelectrochemical hydrogen production on α-Fe$_2$O$_3$ (0001): Insights from theory and experiments. ChemSusChem 2014;7:162−71.

[35] Suzuki S. Syntheses and applications of carbon nanotubes and their composites. Intech Pub. ISBN 978−953−51−1125−2.

[36] Fowler JD, Allen MJ, Tung VC, Yang Y, Kaner RB, Weiller BH. Practical chemical sensors from chemically derived graphene. ACS Nano 2009;3:301−6.

[37] Rout CS, Kumar A, Fisher TS, Gautum UK, Bando Y, Goldberg D. Synthesis of chemically bonded CNT−graphene heterostructure arrays. RSC Adv 2012;2:8250−3.

[38] Dulrkop T, Getty SA, Cobas E, Fuhrer MS. Extraordinary mobility in semiconducting carbon nanotubes. Nano Lett 2004;4:35−9.

[39] Singh H, Bhagwat S, Jouen S, Lefez B, Athawale A, Hannoyer B, et al. Elucidation of the role of hexamine and other precursors in the formation of magnetite nanorods and their stoichiometry. Phys Chem Chem Phys 2010;12:3246−53.

[40] Woo K, Lee H, Ahn JP, Park Y. Sol-gel mediated synthesis of Fe^{2}O^3 nanorods. Adv Mater 2003;15:1761−4.

[41] Meng F, Li J, Cushing S, Bright J, Zhi M, Rowley J, et al. Photocatalytic water oxidation by hematite/reduced graphene oxide composites. ACS Catal 2013;3:746−51.

[42] Edwards R, Coleman K. Graphene synthesis: relationship to applications. Nanoscale 2013;5:38−51.

[43] Duret A, Gratzel M. Visible light induced water oxidation on mesoscopic α-Fe^{2}O^3 films made by ultrasonic spray pyrolysis. J Phys Chem B 2005;109:17184−91.

[44] Fujishima A, Zhang X, Tryk DA. Heterogeneous photocatalysis: From water photolysis to applications in environmental cleanup. Int J Hydrogen Energy 2007;32:2664−72.

[45] Kubacka A, Garcia MF, Colon G. Advanced nanoarchitectures for solar photocatalytic applications. Chem Rev 2012;112:1555−614.

[46] Chen D, Ye J. Hierarchical WO$_3$ hollow shells: dendrite, sphere, dumbbell, and their photocatalytic properties. Adv Funct Mater 2008;18:1922−8.

[47] Sivula K, Formal FL, Gratzel M. Solar water splitting: progress using hematite (α-Fe$_2$O$_3$) photoelectrodes. ChemSusChem 2012;4:432−49.

[48] Hou W, Cronin SB. A review of surface plasmon resonance-enhanced photocatalysis. Adv Funct Mater 2012;23:1612−19.

[49] Wang X, Maeda K, Thomas A, Takanabe K, Xin G, Carlsson JM, et al. A metal-free polymeric photocatalyst for hydrogen production from water under visible light. Nat Mater 2009;8:76−80.

[50] Su DS, Zhang J, Frank B, Thomas A, Wang X, Paraknowitsch J, et al. Metal-free heterogeneous catalysis for sustainable chemistry. ChemSusChem 2010;3:169−80.

[51] Dong F, Sun Y, Wu L, Fu M, Wu Z. Facile transformation of low cost thiourea into nitrogen-rich graphitic carbon nitride nanocatalyst with high visible light photocatalytic performance. Catal Sci Technol 2012;2:1332−5.

[52] Tian Y, Chang B, Lu J, Fu J, Xi F, Dong X. Hydrothermal synthesis of graphitic carbon nitride−Bi$_2$WO$_6$ heterojunctions with enhanced visible light photocatalytic activities. ACS Appl Mater Inter 2013;5:7079−85.

[53] Zhang J, Zhang M, Lin S, Fu X, Wang X. Molecular doping of carbon nitride photocatalysts with tunable bandgap and enhanced activity. J Catal 2014;310:24–30.

[54] Cheng N, Tian J, Liu Q, Ge C, Qusti AH, Asiri AM, et al. Au-nanoparticle-loaded graphitic carbon nitride nanosheets: green photocatalytic synthesis and application toward the degradation of organic pollutants. ACS Appl Mater Inter 2013;5:6815–19.

[55] Liu J, Zhang T, Wang Z, Dawson G, Chen W. Simple pyrolysis of urea into graphitic carbon nitride with recyclable adsorption and photocatalytic activity. J Mater Chem 2011;21:14398–401.

[56] Zhang Y, Liu J, Wu G, Chen W. Porous graphitic carbon nitride synthesized via direct polymerization of urea for efficient sunlight-driven photocatalytic hydrogen production. Nanoscale 2012;4:5300–3.

[57] Yan SC, Li ZS, Zou ZG. Photodegradation performance of g-C3N4 fabricated by directly heating melamine. Langmuir 2009;25:10397–401.

[58] Yeh TF, Cihlar J, Chang CY, Cheng C, Teng H. Roles of graphene oxide in photocatalytic water splitting. Mater Today 2013;16:78–84.

[59] Su F, Mathew SC, Lipner G, Fu X, Antonietti M, Blechert S, et al. mpg-C3N4-catalyzed selective oxidation of alcohols using O2 and visible light. J Am Chem Soc 2010;132:16299–301.

CHAPTER 4

Conclusions and New Directions

4.1 CONCLUSIONS

Semiconductor nanostructure based on heterogeneous photocatalysts have facilitated the rapid progress in enhancing photocatalytic efficiency under visible light irradiation, increasing the prospect of using sunlight for environmental and energy applications such as wastewater treatment, water splitting, and carbon dioxide reduction. Until now, diverse nanostructures of semiconductor have been synthesized and combined to design excellent heterogeneous nanocomposites, which are active under solar irradiation. In view of this, Chapter 1 begins with the general introduction of the subject semiconductor heterojunctions for photocatalysis applications. Initially, we explained different types of advanced oxidation processes and principle of photocatalysis. After this, we summarized basis parameters required to design efficient photocatalysts and reaction steps involved during the degradation of organic pollutants under irradiation. With this, we mentioned the present different photocatalyst junction types such as homojunction, metal/semiconductor, and semiconductor/semiconductor. Further, semiconductor nanocomposites showed higher absorption of solar energy, and to degrade contaminants, formation of heterojunctions with metal and carbonaceous material has synergistic effect toward the separation and transportation of charge carriers, and sensitization of quantum dots of different semiconductors/metals enhanced the utilization of sunlight or improved the photocatalytic performance. The achieved progress in the synthesis of nanocomposites affords a promising route to enhance the photocatalytic efficiencies of photocatalytic semiconductors. In Chapter 2, we briefly explain about nanostructured materials and their interesting properties for the development of advanced devices in different fields. It has been reported that nanostructured materials have high specific surface area, exhibit quantum size effect resulting in an excellent absorption under UV and VIS regions of solar spectrum. The superior properties of nanostructures are useful in advanced functional device fabrication.

Moreover, we included applications of photocatalysis such as water and air purifications, self-cleaning and degradation of different types of bacteria. Finally Chapter 3 deals with the different semiconductors such as ZnO, CdS, Fe2O3, g-C3N4, and their composite with graphene and CNTs for water purification application. These nanocomposites have been successfully synthesized using chemical methods, such as precursor sintering, chemical bath deposition, sonication, and aqueous chemical at low temperature. The structural, optical, and morphological features of these nanocomposites are strongly dependent on preparative parameters, such as pH of solution, precursor concentration, growth temperature, and time. The synthesized photocatalysts have been used to degrade different contaminants present in water. This improvement in performance is attributed to the effective separation of charge carriers, the plasmon-enhanced absorption of visible light, and the large specific surface area for the adsorption of RhB. Photoelectrochemical measurements showed that the ternary hybrid resulted in rapid charge transport and long lifetimes of photogenerated electrons compared with that of binary hybrid. Furthermore, the ternary catalyst was highly stable and could be reused after four consecutive cycles without any noticeable change in the photocatalytic performance. We demonstrated that the incorporation of graphene, CNTs, and other semiconductors resulted in superior photocatalysis performance with potential applications in water purification and large-scale environmental remediation, and this can be achieved using low-cost methods. From various nanocomposites, it was found that introduction of various conducting materials and semiconductors have played a much larger role in this photocatalysis compared to other semiconductor photocatalysts due to its cost effectiveness, inert nature, and photostability. The resulting combined processes revealed a flexible line of action for wastewater treatment technologies. The choice of treatment method usually depends on the composition of wastewater. It was seen that fabricated photocatalysts in this work are efficient and could be applicable for the fabrication of water purification devices. However, a lot more is needed from engineering design and modeling for successful application of the laboratory-scale techniques to large-scale operation. Finally, we addressed scope and development of photocatalytic degradation of organic contaminants, important parameters regarding fundamental principles, and application in water purification under solar irradiation.

4.2 NEW DIRECTIONS

In this perspective, we summarized briefly about various approaches reported in the literature about heterogeneous photocatalysis nanocomposites for the application of water purification. It has been seen that photocatalysis is a green energy source to provide energy and environmental remediation. Until now, significant development in fundamental as well as technological research in the last two decades—photocatalysis based on heterogeneous nanocomposites—showed encouraging properties and could solve energy and environmental problems. Many factors govern the photocatalytic activity of nanocomposites. For an efficient photocatalyst, it should have optimum band gap energy to absorb solar light, minimum recombination of electron–hole pairs, effective separation and transportation of charge carries, high surface area, desirable position of conduction band and valence band, high chemical stability, and capability to degrade wide range of contaminants. Although the photocatalytic processes involve a complicated sequence of multiple synergistic or competing steps, the maximum utilization of solar energy and improvement in separation and transportation of charge carriers are the main challenges and current trend to design highly effective photocatalysts. A variety of routes including doping, surface modification, formation of metal/semiconductor, carbonaceous/semiconductor, and semiconductor/semiconductor heterojunctions were studied substantially to enhance the efficiency of photocatalytic activities. The fabrication of semiconductor/semiconductor heterogeneous nanocomposites demonstrates its perfect photocatalytic effectiveness through utilization of sunlight, improvement of the separation/transportation of the photogenerated electron–hole pairs, and creation of sufficient built-in potential for redox reactions. Although tremendous progresses have been achieved in the investigation of heterogeneous nanocomposites, there are still some challenges in designing high superior photocatalytic systems. Still these nanocomposites suffer various problems such as there is no perfect and detailed understanding of the charge generation, separation, and transportation across heterojunction of photocatalysts, which are important for the design and optimization of highly efficient nanocomposite photocatalysts. Additionally, most of the nanocomposite photocatalysts are UV or near UV active, which reduce the generation of charge carriers greatly. Finally, chemical stability and recyclability of nanocomposite photocatalysts are one of the major challenges to design practical devices. Therefore, the present infancy stage of new century future trends for development would include:

i. Synthesis of nanocomposite photocatalysts, which would be capable of selective photocatalytic degradation of organic pollutants.
ii. Development of cost-effective and novel synthesis methods for ternary nanocomposites for effective degradation.
iii. Detailed theoretical investigation on charge generation, separation, and transportation with the development of new nanocomposites.
iv. Design of more reliable photocatalyst that can cover maximum solar spectrum and high chemical stability.

Printed in the United States
By Bookmasters